Multilingual Multimedia
bridging the language barrier
with intelligent systems

Edited by
Masoud Yazdani

Department of Computer Science
University of Exeter

OXFORD, ENGLAND

First published in Great Britain in 1993 by
Intellect Books
Suite 2, 108/110 London Road, Oxford OX3 9AW

Copy editor: Cate Foster
Cover design: Stuart Mealing
Text layout: Mark Lewis

British Library Cataloguing in Publication Data

Multilingual Multi Media : Bridging the
Language Barrier with Intelligent Systems
 I. Yazdani, Masoud
 410.285

ISBN   1-871516-30-7

Printed and bound in Great Britain by Cromwell Press, Wiltshire

CONTENTS

PREFACE

In post-1992 Europe, knowledge of at least one other European partner's language can not be dispensed with. To prepare for this, it is necessary to look for new strategies which would enable people to communicate with each other in the other person's tongue.

At the moment electronic mail systems can be used between people who understand each other's language. However, for some time we have been concerned with the possibility of a multilingual communication system. Our aim is to encourage users to communicate electronically with speakers of other languages. The growth of the use of various European and international email (and fax) systems indicates that our service would be attractive to many potential users who may need to speak many different languages.

Our earlier work (reported in Chapters 5-9) has attempted to support the user with various Artificial Intelligence (AI) linguistic tools. Our LINGER grammar analyser, checks, offers advice and corrects users' grammatical errors in a variety of European languages. This work has shown us the difficulty of providing effective tools which "know" about a variety of languages and may support a variety of users with different linguistic backgrounds.

As we believe that languages are learned and not taught, we constructed an environment around our system within which the novices could be motivated to learn through their own initiative. Chapter 1 describes *Restaurant*, a prototype system developed in HyperCard for practising foreign languages in a restaurant situation. It incorporates text, animated graphics and digitised speech, and is therefore a sophisticated interactive restaurant phrase book. In addition, it may be used as an interface to the LINGER language-processing system described later in the book.

Chapter 2 presents a survey of approaches to the development of multimedia software and relevant multilingual issues. It gives a definition of multilinuality and moves on to present a framework for a multilingual multimedia database of learning materials. In addition, the article considers the multicultural aspects of such a database, identifying problems which need to be dealt with in this context.

How could two people who do not know each other's language communicate? People do not just use words to communicate with each other. However, studies of the impact of computers deal mainly with linguistic communication via electronic means. We are building an iconic counter-part to a "word processor". Chapter 3 presents design consideration for a visual language and how it could be developed to such a level as to enable presentation of factual information in a way that most people could understand without much difficulty the meaning of the message without prior training.

This iconic communication system uses icons which represent units of meaning greater than single concepts. In return the icons can explain themselves in order to clarify the meaning and provide the context. Such "self-explaining icons" use simple animations to help the user understand the meaning of the message clearly and thus avoid the problem of ambiguity associated with static icons.

Chapter 4 presents a prototype hotel-booking system developed in HyperCard which uses icons to allow a potential guest and hotel manager to communicate. The system is an initial attempt to create an interactive, iconic dialogue using hotel booking as theme. This prototype system would allow a user to compose his complete booking requirement iconically and send the message to the hotel manager to reply to (again iconically).

Chapter 5 describes a system for the teaching of Modern Languages that incorporates "human-like" knowledge of the domain that it is teaching. The project is a progression from the previously developed method of constructing tutoring systems for specific tasks towards producing tools capable of greater generality and use. We describe LINGER (a language-independent 'bug finder') in the context of the experiences gained when we tried to extend it to deal with the Spanish language.

Chapter 6 presents a critical evaluation of LINGER carried out at Purdue University, USA. The article sees the role of LINGER as a prototype of a "linguistic calculator" useful for a variety of applications. A system with such flexibility in use would be a dual-purpose learning environment, one in which the instructor can have his way, provide his best guidance, but hopefully one in which the student can also act in a powerful and positive way to correct and augment the system itself and through doing so develop his own knowledge as much as he cares to.

Chapter 7 describes enhanced LINGER (eL), a system for automatic syntactic error correction intended for use in teaching English to adult learners. The chapter explains how the system uses certain Artificial Intelligence techniques, in syntactic error analysis and correction. We also explore how these novel ideas can be combined with new insights in user interface design to produce software which has an application in language learning.

Chapter 8 offers an assessment of the performance and architectures of a number of programs that are intended for use as grammar checkers. Seven programs are examined to varying degrees - Correct Grammar for Windows, CorrecText, Reader, PowerEdit, Right Writer for Windows, Grammatik for Windows and LINGER in terms of both performance and, where permissible, architecture. It is concluded that none of the programs could be offered to a language learning student with any degree of confidence that their use would improve the student's competence or performance, as mostly they do not address adequately the central problem of syntactic representation of a sentence. That said, when compared against the criteria of grammatical coverage and accuracy, it is clear that the programs fail in different ways and that the architecture and performance of LINGER, although themselves deficient, point the way to what such a program should do.

Chapter 9 presents a overview of the projects which have attempted to apply Artificial Intelligence to the task of teaching a second language. FROG, a French robust natural language parsing system and others which have attempted to remedy its shortcomings are critically evaluated. It is argued that any successful system development needs to be based on sound theoretical foundation. A working hypothesis, L* is presented as a candidate for a theory of how people learn a second language by relating it to their mother tongue.

Acknowledgments

The work reported in this book has enjoyed the financial support of SERC, ESRC, the Training Agency (formerly MSC), German and Italian state telephone companies, the US Army Research Institute and the EC's DELTA project. In addition to those whose contribution is acknowledged in the authorship of the various chapters I need to thank Jonathan Barchan, Gordon Byron, Keith Cameron, Dorian Goring, Steven Jones and Judith Wusteman who contributed to the work of the group.

Masoud Yazdani
Exeter, January 1993

How to claim your free software

Purchasers of this book are offered free software programs, including HyperCard stacks and animated icons, all for the Apple Macintosh.

To claim you free software please complete and return the form below.

Please send me the free HyperCard stacks and animated icons.

Name (Mr/Ms/Dr)
Organisation
Address

Post Code   Country

Date Book Purchased

Place of Purchase

Return to: Intellect Books, 108-110 London Road, Headington, Oxford OX3 9AW, England.   *MM/93*

CHAPTER 1

A multilingual multimedia restaurant scenario

David Pollard and Masoud Yazdani

This chapter describes a prototype system for use by foreign language learners in the practice of language functions and vocabulary related to interactions between restaurant staff and their customers. The program was developed in Apple's HyperCard. "Restaurant" has been developed so as to be usable on its own, but work is underway to link it to Exeter's "eL" natural language parser, described in Chapter 7.

When used stand-alone, "Restaurant" is in essence a sophisticated restaurant phrase book which integrates text with animation and sound, attempting to place the user 'in situ'. By giving the target phrases an interactive multimedia context their exposition is rendered more meaningful.

This work was carried out in the context of a DELTA project dealing with Multilingual Aspects of a Multimedia Database of Learning Materials (Olimpo et al., 1989), described in Chapter 2. One of our aims was to develop a core program which could be used for the learning of a number of European languages. In so doing we were hoping to discover the advantages and problems of using the same multimedia learning sequences for different purposes where the culture and language of users may be different from each other.

Description of a learner session

"Restaurant" begins by presenting an introductory screen (Fig 1) which allows the user to control a number of parameters affecting the course of the following session. At present these choices include the target language (English, French or Spanish), whether the interaction has sound or not, and whether the waiter is polite or rude. The intention is to add further parameters determining such factors as how many customers there are, how expensive or clean the restaurant is, or how good the food is.

Fig 1
The Introductory screen gives the user control of parameters affecting the course of the session.

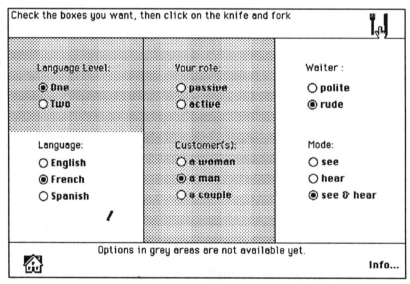

The user then follows a route through a number of interactions between customer/s and restaurant staff in a typical restaurant scenario. There is more than one possible route: the particular route taken during a session depends on choices made by the user either on the introductory screen or during the restaurant dialogue. Thus a user may use the program a number of times in order to examine how his choices have affected the dialogue.

When the user has selected options from those available on the introductory screen an animation sequence (Fig 2) takes the customer into the restaurant.

Fig 2
Having selected options from the introductory screen the customer enters the restaurant.

Once inside the restaurant, the learner controls the action by selecting icons which appear at the top of the screen. In Fig 3, one must choose which character is to speak next.

The icons at the bottom of the screen allow one to quit, to return to the beginning of the program, to seek help (not yet implemented), or to swap between how one wants to be presented with the dialogue and in which language the presentation is to be. These icons remain constant whereas those in the upper control panel continually change.

Let us assume the user selects the customer icon. As the user has chosen to see and hear the dialogue, she first hears the customer speak, then is presented with the equivalent text as shown in Fig 4.

Fig 3
Having entered the restaurant the user chooses which character is to speak first.

Fig 4
The user has chosen to see and hear the dialogue and is presented with the speech text.

Note that a dictionary icon has appeared at the bottom of the screen. The user may now look up a dictionary definition of any word appearing in the speech bubble by clicking on the word. The definition may include examples, graphics, an animated sequence or pointers to further activities related to the word and its use (see Fig 5).

The phrase may be heard again by clicking on the icon of the customer. It may be heard without the speech bubble being visible by selecting the "hear" option at the bottom of the screen then clicking on the customer icon. The speech bubble may then be reinstated by selecting "see" or "see and hear" followed by the customer.

Fig 5
A dictionary facility allows the user to look up any word appearing in the speech bubble and receive a definition which might include examples, graphics, animated sequences and pointers to further activities.

When the user wishes to move to the next stage of the dialogue, she selects the icon of the pointing hand at the top left of the screen. The waiter then replies as determined by the polite/rude option selected on the introductory screen. In Fig 6 the waiter is rude.

Fig 6
The waiter is rude.

Fig 7
A Spanish
equivalent of Fig 6.

English and Spanish phrases performing the same linguistic function can be heard and/or seen by selecting the relevant language options in the lower control panel. Fig 7 shows a Spanish equivalent to Fig 6.

In Fig 8 the learner has the choice of directing the session along two different routes. Selecting one of the icons at the top of the screen decides whether the customer has booked a table or not. This choice will then affect the following interaction and the language used in it. If a reservation has been made, for example, the waiter will have to ascertain the customer's name.

Fig 8
The user directs the flow of the session by deciding whether the customer has booked a table.

In the manner described above, a learner could follow the scenario through the stages of ordering, eating and paying for a meal. At present, however, only the foyer scene has been implemented.

The end of a session

The main educational value of the program lies in the time a learner spends working through the scenario (described earlier). In the process, the learner will be exposed to meaningful use of restaurant language at her own pace. She will have the opportunity to practice pronunciation, and check on vocabulary and grammar (not yet implemented) whilst choosing her own route through the scenes. Hopefully she will work through the scenario more than once, setting up varied conditions at the beginning and observing the changes in language which they generate.

Fig 9
The complete dialogue can be seen, swithcing between languages if desired.

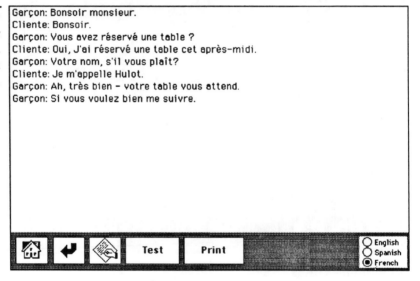

```
Garçon: Bonsoir monsieur.
Cliente: Bonsoir.
Garçon: Vous avez réservé une table ?
Cliente: Oui, J'ai réservé une table cet après-midi.
Garçon: Votre nom, s'il vous plaît?
Cliente: Je m'appelle Hulot.
Garçon: Ah, très bien - votre table vous attend.
Garçon: Si vous voulez bien me suivre.
```

Test   Print

○ English
○ Spanish
◉ French

A number of optional follow-up tasks based on the dialogue may be performed at the end of a scene in the scenario or at the end of a session. At present there are two options which serve merely as examples:

(i) Viewing the complete dialogue

The learner has the opportunity to see the whole dialogue, switching between languages if so desired (Fig 9). When switching between languages the user will not see exact translations, but language performing roughly the same function.

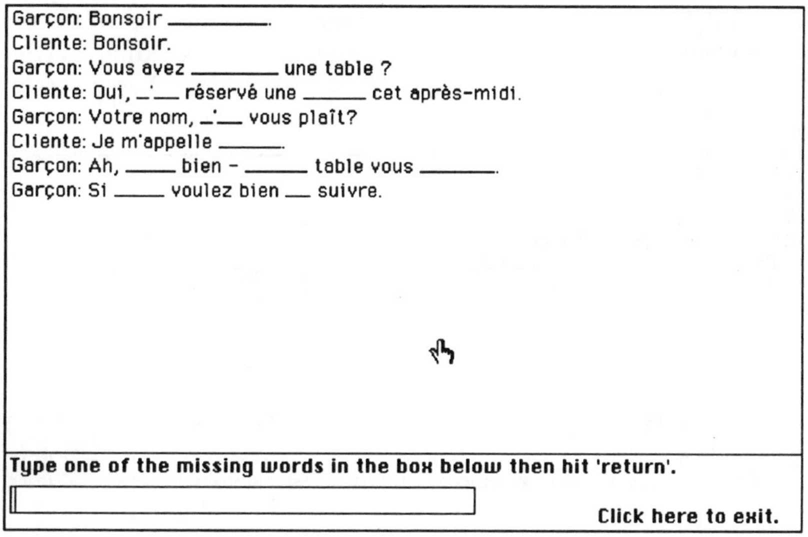

```
Garçon: Bonsoir _____.
Cliente: Bonsoir.
Garçon: Vous avez _____ une table ?
Cliente: Oui, _'__ réservé une _____ cet après-midi.
Garçon: Votre nom, _'__ vous plaît?
Cliente: Je m'appelle _____.
Garçon: Ah, _____ bien - _____ table vous _____.
Garçon: Si _____ voulez bien __ suivre.




                              ☞


Type one of the missing words in the box below then hit 'return'.
[                                        ]
                              Click here to exit.
```

Fig 10
The text of the dialogue may be printed and a cloze-style exercise performed.

He may then print out the text or perform a cloze-style exercise on it as shown in Fig 10.

(ii) Resumé Writing

This option is offered because it is a way of linking "Restaurant" to Exeter University's "eL" parser. It is at present under development and may take a different form when completed. The learner is able to write a resumé of the scenario he has seen by building sentences using words selected from menus of word categories (see Figs 11-13). When a sentence is complete, it can be checked for grammatical mistakes by the "eL" parsing system.

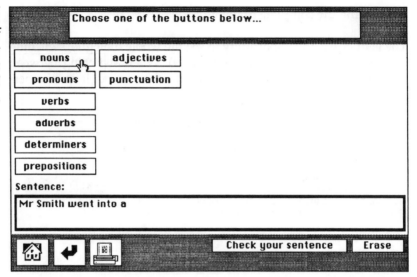

Fig 11
Writing a resume of
the scenario using
words selected from
menus or word
categories.

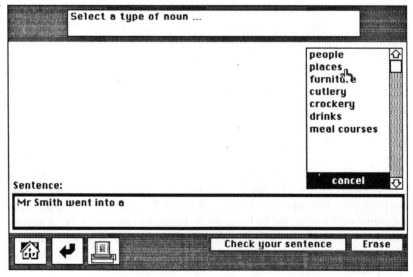

Fig 12
Selecting a noun.

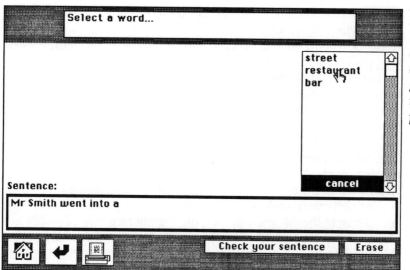

Fig 13
When a sentence is complete, it can be checked for grammatical mistakes by the eL parsing system.

Linking "restaurant" to "eL"

eL is a natural language parser implemented in LPA Prolog on an Apple Macintosh which analyses simple declarative sentences and offers advice on the correction of any grammatical mistakes it finds.

The manner of linking "Restaurant" to eL described above was chosen because the sentence patterns and vocabulary with which eL can cope are limited at present. The structure of sentences input to eL therefore need to be limited so that the parsing system may function successfully. Forcing the learner to create his sentences by a series of menu selections imposes the required limitation on vocabulary and structures. It remains to be seen whether such limitations will intrigue and challenge or whether they will simply frustrate the user.

Concluding remarks

The system at present covers a small portion of a full restaurant scenario, but shows how such a scenario would work when fully developed to include such scenes as sitting at a table, ordering, paying and leaving the restaurant. As it stands it has met with positive feedback when demonstrated and when fully developed will generate long periods of use by learners' self-motivated exploration.

The flexibility of the system is one of its strengths - the user has a large amount of control over what happens during a learning session. The cartoons linked with "real" sampled voices are appealing, as is the humour generated by the rude waiter. User interaction has, however, not been analysed.

On-screen instructions for using the system ("Type in one of the missing words..." etc.) are only in English. These should be available in whichever of the system's three languages the user desires.

The dictionary look-up facility at present only works on single words due to the difficulty of lumping together a group of words in a HyperCard (version 1.2.5) field. HyperCard version 2.0 has overcome this problem.

Ways of making the scenario authorable have been considered, allowing a user to insert easily a new branch or piece of dialogue at a particular point. This is, however, not practical, as any addition of material will need a corresponding sampled voice, icon or graphics which would not match the existing material.

The link to eL is at present a tag-on to the system which serves to prove that linking between a HyperCard stack and such a parser is possible. However, a way of using the parser has to be found in which the parser plays a more central role.

There are a number of ways in which the system could be developed. To begin with one should analyse how the current system copes with learners of different backgrounds, and lessons learned could then be used in developing a final interface design.

The scenario should offer different levels of language difficulty and so it will be necessary to classify vocabulary, structures and idioms to be included at each level. Thus the user will able to choose between different levels of sophistication of the same language.

In order to complete the restaurant scenario beyond the current stage, one needs to draw up a storyboard for the full restaurant scenario, including all branches and possible conditions to be manipulated by the user. This would have to be completed fully to allow the consistent recording of the voices.

To support the system one should design and create a full suite of multimedia HyperCard dictionaries and grammar/vocabulary activity stacks intended to be accessed not only by this program, but by any other HyperCard CALL stack.

The above considerations are a prerequisite to the implementation of a final version which will include recoding. The existing code is to some extent "hacked", many changes having been required to take account of needs and ideas which arose during the prototype stack's development. Restructuring is necessary for optimal performance.

The present graphics were created by editing scanned versions of a particular artist's published cartoons. Permission would need to be obtained to continue to use this method or else bespoke images would have to be commissioned.

Acknowledgments

Thanks go to Gordon Byron, Stephen Jones and Dorian Goring for their help with the work described in this article. This work was funded through a grant from the EEC DELTA programme.

References

Olimpo, G. *et al.* **1989**
   Educational Systems Based on Multimedia Databases
   Istituto Per Le Tecnologie Didattiche CNR (ITD/CNR)

CHAPTER 2

Multilingual aspects of a multimedia database of learning materials

Masoud Yazdani and David Pollard

The use of computers in education has been dominated by a debate between those who support a teacher-directed style through the production of focused courseware (Sleeman and Brown, 1982; O'Shea and Self, 1983) and those who favour production of open-ended learning environments (Papert, 1980, Lawler 1984). The former are effective for the teaching within narrow domains but are at a loss when dealing with the teaching of general problem-solving skills. On the other hand, learning environments, whilst increasing the pupil's motivation, may simultaneously decrease learning efficiency. The pupil simply may not be very good at selecting his own learning strategy.

A sensible solution is to provide resources which can be used by either of these two methods. Such a Database of Learning Materials (DBLM) may in fact encourage new ways of learning to emerge. One of these, guided discovery learning, attempts to allow the learner to explore the material in a free way and be monitored by a guide who at appropriate times may interject helpful suggestions. In fact if a DBLM could be built such that it has no commitment to a particular style of instruction, a multipurpose DBLM is produced.

In order for DBLM to reach a satisfactory level it also needs to be a multimedia resource. Finally, it needs to be free from any cultural bias so that the same material may be adaptable to different languages. These three Ms (Multipurpose, Multimedia and Multilingual) are the definition of an M-DBLM.

14

Building a database of learning materials

What makes a collection of presentations on a topic effective is the care with which it has been put together. A teacher identifies what she wants to teach and works out how it could best be taught to the intended audience. However, this means that the same material would not be much use to another teacher who may want to teach a different topic or even the same topic in a different way.

The Collection of learning material in form of Hypertext (McAleese, 1989) is becoming a popular learning resource where the learners are expected to explore the material with little guidance. This has led to the 'lost in hyperspace' syndrome. The user does not know where interesting things are and what their significance is to her learning goals.

A solution to this problem is to build a database of learning materials (DBLM) which can be used for open consultation as *well* as for guided navigation and possibly as a resource for authors of CAL packages. In this way the learner has the freedom to explore and the choice to use a guide if she wants to. In addition a teacher who may want to build a new course may use portions of the DBLM and exploit it as a general resource.

This approach requires a clear conceptual design for the database such that the same material could be put to different uses. In order to do this we propose that a Unit of Learning Material (ULM) should consist of two parts: the objective data in audio, video, graphics etc., and their educational function within a particular style of teaching.

The separation of the basic data, such as a 3-minute video clip from its educational function (the subjective description of it) is significant for our attempt at making it possible for the same material to be used for different purposes. We call the raw data a Multimedia Sequence (MS) and its educational function an Envelope.

The production of a DBLM relies on the use of facilities such as editors of various kinds for putting together various bits of data in a way that the result conveys the message intended.

In addition we need to allow for a ULM to be produced not by a human developer, but through a computer program such as a simulation or an intelligent tutoring system (ITS). On top of that we need to allow for the fact that ULMs of different kind may need to be put together into Learning Sequences (LS) which themselves have a higher level objective than the objectives of the individual ULMs.

Let us try to write a more formal definition of a DBLM.

DBLM -> Root + {View}*
Root -> {LS}*
View -> {LS}* + Envelope
LS -> {ULM}* + Envelope
ULM -> Program + Envelope
ULM -> MS + Envelope
MS -> text, data, graphics, still images, animation, video, audio.
Envelope -> Intention + Facility
Facility -> Editor, Programming Language etc.

In order for a DBLM to be useful it needs to contain some specific examples of ways in which it could be used. A specific courseware on a topic within DBLM can be called a View. A M-DBLM needs to include a number of off-the-shelf views in addition to facilities for the users to build their own personalised views of the DBLM. In order not to duplicate material we may decide to keep the core of the material in a DBLM in a Root directory and allow the rest of the Views to be defined as modifications of that Root directory.

A framework for a multilingual DBLM

Computers are being used by people all over the world. This universal use of computers poses new and interesting problems for software developers.

Most computers and the software that controls their functioning are produced as if the users were native speakers of English. However, these systems are used by people who may not know any English. Although cosmetic changes are widespread (such as special keyboards) the problem needs major attention. There are three aspects to the problem:

1. How to make it easy for non-native speakers of English to use computing systems originated in English-speaking countries.

2. How to make computing systems produced in the rest of the world useful for English speakers.

3. How to produce software which is without linguistic or cultural bias inherent in its structure.

Using a computer system is emotionally charged even without having to worry about language problems. Trying to learn from a computer system which is not in the user's mother tongue is even more difficult.

The problem arises at many different levels. The simplest level is lexical, the words which are used to interact with the system. These can be changed reasonably easily into a new language. However, this change would give a false sense of security to the user. The underlying syntactic and semantic structure of the system may still be organised in such a way as to confuse the users. An Italian may see Italian words used to describe objects stored in a database, but the structure of that database in most cases would be based on the conceptual structure of how the English people and English language see them related.

In most cases those struggling to translate software into other languages find themselves in dire straits. Some superficial aspects of a system are changed but more important aspects remain in the original language. The user does not know if her confusion is due to unfamiliarity with the system or due to language problems.

The European nature of the DBLM project means that people with different languages will be building and using the outcome of the project. How is this multilingual context to be dealt with? We shall look at two issues relating to the development of the system (Are the DBLM materials to be produced in different languages?) and its use (Is the mother tongue of the users different?).

One simple solution is to have multiple DBLMs in different languages, very much like books which are translated into different languages. This seems inappropriate, as a DBLM is a system which may need updating more often than a book.

At the other end of the spectrum one may dream of a day when machine translation facilities may translate the DBLM into different languages at the user's request. The option at this moment is not open to us.

What seems more realistic is that the DBLM should be in different languages where some bits may be in more than one language. The users, on the other hand cannot be expected to be multilingual.

- We propose that the DBLM should be defined in such a way that language is an attribute of the system. Depending on the user's language we show that person the parts of the system available in that language. These will present different views of the DBLM.

- In addition all facilities of the DBLM need to have multilingual dialogue boxes. So, depending on the user's preferred language different menus and dialogue boxes relating to system utilities are shown.

- Finally, it seems important that the user is provided with language-specific browsing facilities. The system builds a language specific view of the database for the user with those aspects which are excluded in shaded form (as in the Apple interface). This way the user is basically saying "Show me the Italian bits but also tell me what is there that I am missing?".

- It should also be possible for a user to switch to another language while using the system in order to see non-language specific material excluded from her chosen view.

This is only an incomplete list. The following sections present one of the options that we have considered in more detail.

Multilinguality and multilingualism

In our project we have found the definitions by Hamers and Blanc (1989), have proved useful for the designer of a DBLM.

Multilinguality can be defined as "the psychological state of an individual who has access to more than one linguistic code as a means of communication" whilst *multilingualism* is "the state of a linguistic community in which a number of languages are in contact with the result that different codes can be used to perform the same interaction".

A DBLM intended for Europe-wide use must possess a multilingualism which allows the achievement of goals in different languages. User goals will include the construction of ULMs and learning sequences by pedagogues and learner browsing the system or following a learning sequence. Linguistic knowledge of a user must not , as far as possible, prejudice her access to the DBLM and its facilities.

The database must also take into account the linguality of its user, allowing and encouraging her to make use of her multilinguality if she is master of more than one language, but not diminishing her access if she is monolingual.

Facilities

Facilities must be available in all languages. The mass translation of all elements could be lessened through the use of internationally intelligible icons where practical. Research into the difficulties involved in icon design is being carried out (Yazdani and Goring,1990).

Before entering a facility, a user needs to state one preferred language of interaction in facilities. She must also indicate her competence in any other languages, which will affect the system's presentation of ULMs (see Section 4.2). The user might indicate which languages she knows by selecting from a list of languages used in the DBLM. This might be done by positively selecting those desired or by eliminating those not wanted.

It might be advisable for the user to qualify such a choice of languages by skill or by level. ULMs or elements of ULMs would correspondingly be tagged by level. The user might then be notified if a ULM contains elements of greater linguistic difficulty than that which the user has declared to have mastered.

ULMs

The European nature of the DBLM project means that people with different languages will be building and using the outcome of the project.

As outlined in the conceptual model (Chioccariello *et al.*, 1990) the DBLM must give a European user the possibility of easy access to a pool of learning material in a variety of languages with a minimum of problems.

A ULM may refer to different languages, may be language-independent (for instance a picture or a video sequence without language), or a mixture of language-dependent and independent elements. Also, each language-dependent ULM may be represented in the DBLM in a number of different versions corresponding to different languages.

It has been suggested that the DBLM could be given maximum flexibility by not directly including language-dependent components in a composite ULM. The ULM should include a pointer to a set of equivalent components in different languages: a run-time association between language-independent and dependent components. The binding between the composite ULM and a specific language-dependent component should be made at run-time on the basis of the user mother tongue and of the languages the user has declared to know. Thus, a user's specific language skills will give her access to a specific subset of the database.

The binding process would decide how to build a composite ULM on the basis of the language-dependent components actually available in the database and of the priority the user has given to the different languages she understands. If the ULM element were, for example, a quotation, the element's language of origin could also affect the binding process. Thus, if a user has a good understanding of Dutch and a lesser knowledge of Italian, an Italian quotation available in both languages might be offered in both its original form and the language declared to be better understood by the user.

In order to accommodate the creation of a ULM available in different language versions, the method of construction of ULMs will have to take into account possible differences in the length of text and in the duration of pieces of speech.

A language-dependent ULM may contain elements in more than one language and a translated version of some of its parts may not be available, making the understanding of its whole dependent on more than one language. In this case the user who has not declared knowledge of all the necessary languages should not be denied access to the ULM. In this case, should the user be merely notified about the matter and presented with the ULM in any case, or should the user have the option to skip such a ULM ? When such a ULM is part of a learning sequence, it might be advisable to present the ULM in whatever language is available, for the user has the option of using the remote control facility to skip past an element in an unintelligible language.

Multilingual aspects of the browser

The DBLM will possess a browser enabling the student to discover material (in the form of ULMs) relevant to her needs. The design of the browser can be tackled from two angles. On the one hand, the browser may present the user with the structure of the database, allowing her to pursue paths through the structure which she thinks will take her to a desired ULM. On the other hand, the browser might ask the user to define search criteria relevant to her needs, then present a list of found ULMs fulfilling those criteria. Neither of these methods is ideal, for the former requires an understanding of the organisation of the database (and databases in general ?), whilst the latter requires an understanding of how to formulate formal queries. The matter is being researched at present, and the answer may lie somewhere in the middle, possibly with some form of hybrid system. The examples below, however, assume the option of user query formulation.

In order to account for multilinguality and possible monolinguality of a user, after user definition of a query, the search mechanism might only seek out those ULMs available in languages declared as known by the user. This would seem to be too limiting. A search would do better to ignore language availability until the user wishes to "view" a ULM. Thus, the results of a search might be presented to the user as a choice of the languages in which relevant ULMs have been found. Alternatively, the browser could present the user with the titles of all ULMs relevant to the search - translated into a language the user understands and organised by the language/s in which they are available.

If a user is to have access to ULMs in languages she has not declared as to know, then a small amount of information about each ULM must be available in all EEC languages in order to enable the user to judge whether it is worth her while accessing the ULM. As well as the title, such information might include a short description, the communication type, and the language/s it contains. Appendix A shows how this information could be presented by the browsing facility. A further element might indicate the level of linguistic difficulty incorporated in the ULM. This would encourage a user with weak mastery of a language to "view" a ULM tagged as easy to understand, and avoid the disheartenment of a student viewing a linguistically difficult one, for she will at least be aware of the fact that the piece is acknowledged to be difficult and will not be put off viewing all ULMs in that language.

Conclusion

Coping with the demands of a multilingual user base, even within such a culturally close-knit community as the EEC countries, is a non-trivial affair. Language will feature prominently in almost any educational system. The system must cater for both monolingual users and multilingual ones, encouraging users to admit to and use their knowledge of foreign languages, whilst not forcing such use upon them and thereby hindering their attempt to achieve whatever their goal is. The choice must lie with the user.

Multicultural aspects of a DBLM

Defining "culture"

An internationally accessible DBLM must not only take into account the languages of the participating countries but also aspects of culture which may cause difficulties in its creation or use.

"Culture" is a spurious term which may include the language, politics, religion, living conditions, social structure, philosophy, geography, history, economy, dress, music or other art forms of a community. Boundaries which lump people together as belonging to a certain "culture" must be fuzzy and are often indefinable. At a stereotypical level, people of the same country may be said to have the same culture, but doing so might lump together the Basques and the Milanese with the Catalans, the Scottish highland crofter with the London finance broker, Luciano Pavarotti with Sabina. The British politician, Nicholas Ridley, has recently discovered the errors of such closeted thought, for it is treading dangerous ground to refer to culture in terms of the passport a person holds. How, then, should boundaries of culture be drawn - along the lines of class, age, politics or sexuality?

In the authors' view, cultural boundaries are, for the practical purposes of this project, indefinable. However, the question of "culture", or factors one may define as broadly "cultural", does gives rise to debate on a number of issues that need to be addressed by the creators of a DBLM. The crux of the matter is in basis how to deal with subjectivity of all kinds.

These problems arise at two levels: in the creation of Units of Learning Material (ULMs) and in the linking of these into Learning Sequences (LSs).

Cultural aspects of ULMs

A ULM could be created by one person, a group of people in a single workplace, or a dissipated group of collaborators. If a ULM is to have a single creator, that person will have to possess both subject domain knowledge and expertise in using the ULM creating facility. It is therefore more likely that a ULM will be created by a group, although the work of a group will probably be just as subjective as that of an individual. Whether such subjectivity might arise through oversight or premeditation, a number of questions need posing which are examined in the following sections.

Should subjectivity be discouraged?

Should the DBLM contain ULMs which are merely factual, forming a bland opinionless body of "objective" material ? How is objectivity to be achieved ? Are advisory guidelines to be issued to ULM creators ?

Is the aim of the DBLM to create material with a unified philosophy? What form should this philosophy take? Should there be a general promotion of humanitarian values? A capitalist western European view of the world ? A "one world" philosophy?

Should subjectivity be welcomed?

If the DBLM is to act as a free forum for all cultures, is this freedom to be total or is it to be regulated ? Anyone who subscribes to unregulated electronic network newsgroups will have experience of the medium being used to express offensive views. Is such "offensive" material to be allowed in ULMs? Indeed what is offensive - racism, sexism, sex, violence ?

Two pieces of British legislation raise interesting issues:

(i) Arts promoting homosexuality are currently banned in Britain, for the British government sees homosexuality as offensive. However, many of its subjects are offended by homophobic attitudes. This begs the question of how the administrators of a DBLM might reconcile conflicting views of what is acceptable.

(ii) The broadcasting of comments by the political party Sinn Fein are also banned in Britain. This is due to the party's connections with the IRA and terrorism. Ignoring the question of whether one person's terrorist is another's freedom fighter, one must ask whether all DBLM users should be denied access to material which is banned in only one of the partner states.

Censorship

If there is to be regulation of material in a DBLM, who should wield the power of such control? Should a central European body be set up? Should each country have a different view of the DBLM which excludes material its government finds unacceptable? Should individual institutions have censorial power? Should children be allowed access to adult material?

Propaganda

Should advertising be allowed in ULMs? Industry might put a great deal of effort into producing ULMs which promote their products. Should such ULMs be allowed but labelled as advertisements? How would an industrial producer of a ULM feel about the constituent pieces of its ULM being reused by another ULM creator, possibly a rival? Could this lead to the bank of ULMs being swamped in a battle of industrial or political propaganda?

Declaration of source

Should the source of any ULM always be made clear to the user? If so, could the creator of a ULM be (wrongly) held representative of a certain country or section of society ?

Browsing ULMs

It will be possible to browse ULMs by user query formulation on topics of interest, ULMs having been tagged with keywords to be searched by the browsing facility. Here again culture might throw a spanner in the works, for the choice of keywords used to describe a ULM will necessarily be subjective; for example, a ULM assessed by a right-winger as containing nothing of political significance may have political import in a socialist's eyes.

A solution

Can solutions to all the above questions be found which do not render the creation of ULMs impractical? The following solution addresses some problems and aims at simplicity and workability.

It seems clear that there must be some measure of control over the addition of ULMs to the DBLM. General guidelines need to be drawn up by representatives from partner states as to what is acceptable. These guidelines could cover design standards as well as content, and are to be made available to all creators of material.

The vetting of all ULMs before they are added to the DBLM would seem Draconian, would lead to delays between the creation of ULMs and their use, and is therefore impractical. The DBLM would be better monitored by users themselves by means of a complaints procedure through a moderator with the power to ask the creators of a ULM to modify their work.

Cultural aspects of learning sequences

Educational systems at all levels have always been prey to the influence of culture, this fact being reflected to some extent in all national curricula.

Factors affecting educational curricula

The following general factors affecting a country's educational curricula are paraphrased versions of those identified by Faraj (1988):

(a) **Economic.** Education systems must cater to economic realities. If a country is largely agrarian, the education system must provide workers who can contribute to the country's agricultural productivity in the best way. If an industrial country depends on one major industry, that country must educate a segment of its population to manage or to be workers in that industry. Advanced industrialised countries, on the other hand, need multi-tracked education with differentiated curricula.

(b) **Geographical.** Topography often defines the content of curricula, sometimes to the extent of creating unique types of institutes such as colleges for fishing, forestry and maritime sciences. Meanwhile, climate affects the school starting age, the extreme cold of countries such as Norway and Sweden making it unsuitable for children to attend primary schools until a couple of years later than their counterparts in the Mediterranean.

(c) **Philosophical.** Philosophy directly influences education, since it presents a theory of knowledge and relates this theory to the acquisition of skills. For example, Kant's distinction between the realms of science (the physical) and supra-science (the metaphysical) lead to Germany's early move away from the Enlightenment's unitarian study of science by generalists and amateurs who regarded it as revelation and deeistic justification. Germany was the first country to establish a higher education system of specialised training and investigation depending on rigorous logical precision and empirical observation.

(d) **Cultural.** Covered in the above section.

(e) **Political.** Education may be affected by the theories of those holding political power (e.g. Rousseauism or Marxism) or by a temporary condition (e.g. war). In democratic regimes, governments share the role of controlling education to some extent with the population, although all political parties set educational goals intended to benefit their own political aims.

(f) **Religious.** The period of conflict between church and state during the reformation has left three types of relationship between religion and education in Europe. In some countries (as in the former USSR), sectarian groups are excluded from influence in education. In some (e.g. France), two separate systems exist side by side - secular education in state schools and religious education in schools run by the church. In other countries (e.g. UK), governments co-operate with religious denominations in supervising education.

Learning sequences and curricula

The creation of learning sequences (LSs) which are suited to the curricula of a number of countries must therefore take account of all the above factors for each relevant country. The possible combinations of such factors is huge. It is debatable whether the creation of "multi-curricular" learning sequences is feasible. Were such a task to be attempted, the development team would need to include members with an intimate knowledge of each country's curricula. Progress of work within such a team would be slow, and might simply lead to a deadlock, proving the task impossible.

The most feasible solution would seem to be not to attempt to create overtly multi-curricular LSs. Instead, they will tend to be created to fulfil a particular part of a particular curriculum, but may also fit into different educational schemes. An LS should therefore be tagged with information about the various curricula in which it may be used. The conceptual model of the DBLM allows for the creation of a new LS by modification of an existing LS, so different versions can be tailor-made for different curricula. In order to keep this task of modification as simple as possible, it is helpful for the creator of an original LS to be aware of possible use in other curricula and indeed of possible changes in the curriculum for which it is designed.

The creation of different curriculum-specific versions of a particular LS could be regarded as preferable to a single bland multi-purpose LS which fulfils no-one's needs exactly as they might wish. Such a solution would also allow for the use of the DBLM for comparative educational studies.

Summary

The European context of the ESMBASE project (Olimpo *et al.*, 1989) and the need to experiment methodologies and prototypes in different countries requires a multilingual approach. This article examines the multilingual problems facing an internationally accessible educational resource with reference to a planned European Multimedia Database of Learning Materials. It presents a definition of multilinguality and moves on to present a framework for a multilingual Database of Learning Materials. In this context a general specification of multilingual aspects of a Browser system are presented.

In addition, the article considers the multicultural aspects of a Database of Learning Materials. A number of problems are identified which need to be dealt with in this context. Practical suggestions are made for catering for a user base having varied linguistic backgrounds, using the example of a learner browsing facility.

A longer version of this article, which presents a more detailed description of a prototype "animated specification" developed at Exeter University for the EC's DELTA project D1012, is available on request from the authors..

References

Chioccariello, A. *et al.* 1990
A Conceptual Model for a Database of Learning Material
Instituto per le Tecnologie Didattiche CNR
Faraj, A.H. (1988)
International Yearbook of Education Volume XL
UNESCO
Hamers, J. and Blanc, M. 1989
Bilinguality and Bilingualism
Cambridge University Press
Lawler, R. 1984
Designing Computer-Based Microworlds
in Yazdani (ed.) *New Horizons in Educational Computing*
McAleese, R. (ed.) 1989
Hypertext: theory into practice
Ablex/Intellect Books

Olimpo, G. *et al.*, 1989
Educational Systems Based on Multimedia Databases
Istituto Per Le Tecnologie Didattiche CNR (ITD/CNR)
O'Shea, T. and Self, J. 1983
Learning and Teaching with Computers
The Harvester Press
Papert, S. 1980
Mindstorms-Children, Computers, and Powerful Ideas
The Harvester Press/ Basic Books
Sleeman, D. and Brown, J. S. (eds) 1982
Intelligent Tutoring Systems
Academic Press
Yazdani, M. and Goring, D. 1990
Iconic Communication
Department of Computer Science, University of Exeter

CHAPTER 3

A computer-based iconic language

Stuart Mealing and Masoud Yazdani

In *The Hitch-hiker's Guide to the Galaxy* (Adams, 1979) it becomes possible to understand communication in any language by plugging a Babel fish into your ear. The fish automatically translates from any language to any other language, and exemplifies the long-term dream of mankind of being able to communicate across all language barriers. Computer assisted language translation is a tool of Artificial Intelligence (AI) which will eventually help realise this dream, and the vision of Koji Kobayashi (Kobayashi, 1986) - that it will be possible for an English speaker to pick up a telephone and speak with a Japanese in English - is only ahead of its time in the sophistication required of the system. An alternative approach is to use a single, international language, but attempts to popularise Esperanto have met with little success, largely due to the initial effort required to learn it (by a worthwhile number of people). There are, however, existing signs, symbols and icons which are understood internationally and it is the aim of this research to discover the level of subtlety of communication which icons can achieve, and to develop a computer-based iconic language. This will be designed, in the first instance, to operate within a limited domain such as 'hotel booking', where it is common for language difficulties to arise but where a limited and predefined range of ideas is likely to be expressed. A scenario such as this could be treated as having a 'situational script' (Schank and Abelson, 1977).

Developing the language for use on a computer will significantly impact its design, offering a flexibility and 'intelligence' in presentation and interaction which is not available in other media. Whilst the ideal iconic language would require no learning and would be intuitively obvious to use, it is more realistic to expect some explanation to the user to be necessary, and the system employed will seek to minimise this need. A 'friendly' system will encourage discovery through use and the computer will allow, for instance, for the user to interrogate icons to learn (or confirm) their meaning, this explanation coming in a lower level iconic form. The user interface can also be supported at any point by labelling in written or spoken natural language.

Icons

An icon is a symbol representing, or analogous to, the thing it represents (Collins, 1986), it is something which looks as though it could act as a proxy. Huggins (Huggins and Entwisle, 1974) asserts that: "Iconic communication deals mainly with non-verbal communication between human beings by the use of visual signs and representations (such as pictures) that stand for an idea by virtue of resemblance or analogy to it in contrast to symbolic communications where the meaning of a symbol is entirely nominal (such as English text describing a picture)". Pure icons, therefore, rely initially on recall of a previous visual experience on the part of the user (either first- or second-hand) with sufficient particularity to make their use in a particular context clear to him. They may, however, take on the role of symbols in subsequent use, and can be used in conjunction with symbols, i.e. a diagonal bar through an icon can indicate negation. In the context of this research the language will be designed for display on a computer screen, and prototyped initially in HyperCard™.

Criteria for the icons are that they should be:

- graphically clear
- semantically unambiguous
- without cultural, racial or linguistic bias
- adaptable (open to modification to express nuance)
- simple (perhaps created within a 32x32 matrix)

As it is desirable that the system should not rely at any stage on any natural language, an icon whose meaning is not immediately obvious to the user should be able to explain its meaning in terms of more fundamental icons or through diagramming its evolution from its source imagery. It would, for instance, be possible to 'click' on a compound icon (standing for an event or concept having several elements) and have its meaning expressed by several static base icons (standing for those individual elements). Alternatively, a simple animation of the 'actors' in the base icons could explain the meaning of the compound icon, by performing its message or by diagramming its construction. 'Clicking' on those base icons might initiate a simple, but revealing, transformation from a more comprehensive (possibly photographic) image showing the source of the icon, to the icon itself. The use of relevant photographic images as backgrounds to messages could establish the current domain, and the use of colour will be a helpful adjunct but cannot be used to carry essential information that might become lost on a monochrome system.

The icons will therefore comprise several levels. The top level will present the message and will incorporate compound icons and symbols where needed. Querying any icon will initiate a move to the second level, at which base icons will seek to explain the meaning of the top level icon, acting out the meaning if appropriate. A further level could trace the icon's development from suitable photographic references to its source, and at any level natural language could act in support. The natural language which is active would have been chosen by the user on entering the system. In keeping with the intention of pictorialising the complete interface, this selection would be made through identifying the user's country of origin on a map.

Additional weapons in the design armoury are therefore:

- colour, which might be used either expressively or for coding
- movement, which is potentially very powerful (most obviously for dealing with time dependent concepts such as movement) but not to be used lightly
- background, which can be used as a visual cue to the domain.

Visual language

If a picture is worth a thousand words, an icon must be worth at least a sentence! Pictures can communicate more than just their visual contents because a wide range of prior knowledge and experience is brought to bear on the imagery by the viewer, and this can be exploited to enrich the meaning of an icon. Pictorial images are specifically used for communication in many situations, and their interpretation is often subject to the context in which they are found. For example, images on a can of food often illustrate the contents, but a label showing a picture of a smiling child would not be assumed to be an indication of the contents of a can, since eating people is not part of our cultural heritage (Yazdani and Goring, 1990). Similarly, the interpretation of gestures that accompany spoken language can be subject to context, but can often significantly improve understanding, notably in a situation where the spoken word is difficult to hear or to comprehend. Many people speak little of any foreign language but are able to understand and communicate a surprising amount when abroad with intuitive sign language. As well as these multilingual gestures there are many inter-nationally understood symbols (such as arrows to indicate direction and overlaid diagonals to indicate negation), a study of which can usefully be brought to the design of our icons.

Written sign language is pictographic, that of the North American Indian being described as "elemental, basic, logical and largely idio-matic" (Yazdani and Goring, 1990). As it is likely to be written/drawn on a single surface, the whole message is visible at any moment in time and therefore any part of it is open to scrutiny at leisure in the context of the rest. This is very different from gestural language which is serial in nature, only a small part of the message being communicated at any one moment, and attempts to look for context relying on memory. Whilst it is envisaged that the proposed iconic language will employ the advantages of viewing a complete message (or 'page' of a long message), the alternative of having a message unfolding over time could be employed in part. An animated icon can 'act out' simple time-based concepts (such as walking or shrugging) with the advantage that it can be repetitive and incorporated as an element in a continuously visible message.

It is practical to categorise icons (in the manner of Marshall McLuhan) in terms of the 'temperature' of information which they can convey, according to their position on an axis stretching from pictures to symbols, and on their level of animation(Fig 1). (In other contexts, icons made of pictures have been called 'picons' and moving icons have been called 'micons'.) It is not necessarily the case, however, that the hottest presentation is the best, as it may become too specific to illustrate a general case.

Fig 1
Categorising icons
in terms of the
temperature of the
information.

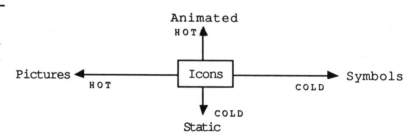

Design considerations

The 'dustbin' icon is now used regularly in GUIs (graphical user interfaces) and provides a suitable example for discussion of some of the considerations necessary in design of an icon. Already we have referred to it as a 'dustbin' icon, whereas in America it would be a 'trashcan' and in France a 'boite à ordures' so it is fulfilling its multi-lingual function, but does it represent a globally used waste disposal method? Even in a Western society, people living in blocks of flats are more likely to use rubbish chutes than dustbins. The icon is also derived from what used to be a typical dustbin, but which is now superseded in many instances by more modern designs. If dustbins cease to be used in the future, subsequent generations will require either a new icon or a short history lesson. Providing, however, that everyone who uses a computer is familiar with that particular form of waste disposal, then it is clear what object it stands for, but otherwise the image will need to be explained. Having understood what the object is, do we now under-stand what function the icon serves in the application? It is not a big conceptual jump from disposing of unwanted material in a dustbin to disposing of unwanted files in an application, but it is one that needs to be explained on first encounter with a GUI, and it is not obvious that the same icon may be used to eject discs in some machines. Typically of icons, the dustbin image is standing for a noun, but is intended to implicitly conjure up a didactic phrase associated with the noun. (As the other icons on the computer screen at the same time as the dustbin are usually standing for items in an office environment, e.g. desktop, files etc., it might be thought that a waste-bin would provide a more appropriate disposal metaphor, and this is sometimes used.) Other, more complex, icons such as a figure ascending a staircase (to indicate an exit route, perhaps), are far more explicit in their description, and have a sentence-level structure.

The style of existing icons falls into three main categories:

- silhouette (in front or side view), either solid shaded or inlinear outline
- three-quarter top view, always linear and usually in isometric perspective
- 'realistic', drawn or semi-photographic (with graded shading

Fig 2
The three main
categories of icons.

etc.)

The icons are normally diagrammatic and may be either combined with, or interspersed with, symbols (arrows, numerals etc.). It is unfortunately apparent that their style is as subject to fashion as any other design product, even typefaces, and that the ideal, timeless range of icons is unlikely to be found. It is, however, possible to avoid the excesses of fashion, and to produce icons which have a long life span and are amenable to future stylistic updating without compromising continuity of recognition. Of the styles described, the three-quarter top view is potentially the most informative but requires a slightly more sophisticated visual understanding, the solid silhouette is the most arresting but somewhat limited in the range of things it can deal with easily, and the 'realistic' is easy to recognise but risks moving too far away from the generalisation which an icon often requires. Pictographic sign language is more calligraphic in its visual character.

It is normally confusing to mix styles within a message, but the possibility of doing so for the sake of emphasis can be explored. The reflex recognition required of a road sign, however, is not the same as that required of icons to be used in the composition of a message. The different styles lend themselves to different degrees of abstraction, and the likely criteria of adopting a set matrix size for the icon will condition the level of detail and stylisation.

Fig 3
An animated icon
may be used once
static icons have
proved inadequate.

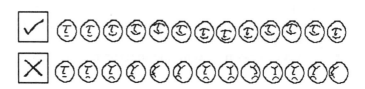

Animation of icons has potential for extending the range of concepts that can be dealt with, but should only be used once static icons have proved inadequate, or as an aid in explaining the meaning of static icons. In the above example (Fig 3) , if the user interogates the 'tick' or 'cross' icon, a head either nods or shakes to explain the symbol (although it should be recognised that these gestures are not universally recognised). It should also be used conservatively to avoid an iconic message deteriorating into a fairground of visual activity. Given those reservations, the possibility of limited animation being used to make connections between static icons, to describe concepts involving verbs, and to be used in the explanation the evolution of any unfamiliar icon, is very promising. Whilst the implementation of the language on computers makes animation viable, storage and display considerations for animated icons must be considered, but current technology and data compression techniques should cope with the relatively limited requirements. The envisaged icons will also lend themselves to vectorisation (now commonly used for fonts in DTP applications) which implies very limited storage, easy transformation and scaling, and a high level of portability. At some future date it might prove relevant to consider the possibility of 3-D icons or of 2-D icons inhabiting 3-dimensional space.

The judicious use of colour can be employed to good effect, most obviously for coding, but the system must be able to communicate fully in monochrome. In many situations, such as at an airport, the location itself acts as a cue to the domain of the icon, but this is not necessarily the case in a computer-based language capable of dealing with many domains. Background images, however, could help to establish context for a message and multi-media developments are bringing closer the option of using video in that role ! In fact, as much multi-media production is visual, and hence alingual, it would be a very appropriate medium in which to use an iconic language.

It is worth noting that the potential standardisation of icons internationally across a range of applications is currently inhibited by legal restraints regarding copyright, and it is useful to consider the implications of an iconic language bridging that barrier.

Syntax

In most everyday situations icons do not interact with their neighbours. They often appear singly and, when grouped, usually constitute a list of features or properties (such as listing the facilities of an hotel). In the earlier example of the 'dustbin' icon, the icon is a passive representation of a word around which the user might build an understanding of its intention. A more dynamic icon would offer a far more explicit instruction about its function, and as such would embody a unit of meaning equivalent to a phrase, sentence, or perhaps even a paragraph. To comprise an efficient language, icons need this flexibility and the ability to deal, for instance, with concepts incorporating the equivalent of natural language verbs. It is unlikely that constructing a message with a one-to-one matching of icons to words would be appropriate, and the size of the unit of meaning which an icon could encompass is likely to grow in proportion to familiarity with the system. The ability to interrogate icons therefore offers the advantages of efficient, high-level, compound icons, with integral explanatory support, the need for which decreases with familiarity. In this context, communicating meaning clearly need not rely on an explicit grammar or syntax and it does not become necessary to think of an icon being translated into natural language as a stage in its comprehension.

Basic icons can be combined into compound icons. For instance, we are familiar with icons for man and stairs being combined to show a man on stairs. An arrow added to indicate the direction upwards produces an icon telling us to use the stairs to go up. This is often combined with a word or an image to explain that the stairs take us to the exit, or perhaps provide our exit route in the case of fire. A diagonal bar through the same icon (often in red) would tell us not to use the stairs. The computer would allow us to produce compound icons with some ease, to modify scale or weight to suggest priority or emphasis, and to create links between icons. Some punctuation marks, such as full-stops and brackets, might also be used.

The establishment of an iconic dialogue is more challenging than the preparation of an iconic message. It could take the form of alternating messages, or the receiver might be empowered to interactively modify, or comment on, the sender's message (perhaps overlaying a request for a hotel room with signs to indicate the availability of the required accommodation on the required dates). Pressure on an icon library would increase with the size of the domain of the dialogue and the situation encourages the ability to develop fresh icons, according to need, from base icons. Comparisons with morphemes and lexemes are possible, but the scale of meaning of a base icon is likely to be greater and open to inflection by modification of its visual detail. Once a dialogue is established, its own dynamic will suggest context and assist with understanding of the icons, but its initiation may require more particular handling.

Conclusion

Icons offer a rich potential for communication across natural language barriers. If confined to the European arena, the many shared conventions (such as the use of Arabic numerals and question marks) make their design much simpler and their sure interpretation more certain. The computer provides an ideal device for the implementation of a flexible iconic communication system.

The need for such a system is made more urgent by the increasingly international nature of commercial, educational and social communication. Examples such as booking a hotel room abroad, ordering machine parts from a foreign subsidiary, or accessing EPOS (the European PTT Open Learning Service) all provide occasions for such a system to prove its worth.

References

Adams D. 1979
  The Hitch-hiker's guide to the Galaxy
  Pan Books Ltd.
Kobayashi K. 1986
  Computers and Communications
  MIT Press
Schank R. & Abelson R. 1977
  Scripts, Plans, Goals and Understanding
  Lawrence Erlbaum Assoc.
The Collins English Dictionary 1986
Huggins W. H. & Entwisle D. R. 1974
  Iconic Communicatio: an annotatedbiography
  The John Hopkins University Press
Yazdani M. & Goring D. 1990
  Iconic Communication
  Dept. of Computer Science, Exeter University

Selected bibliography of icons

Kepes Gyorgy (Ed) 1966
  Sign, Image & Symbol
  Studio Vista, London
Dreyfuss Henry 1972
  Symbol Sourcebook
  McGraw-Hill Book Co.
Diehtelm Walter 1974
  Form + Communication
  Editions ABC, Zurich
McLendon C. & Blackstone M. 1982
  Signage
  McGraw-Hill Book Co.

CHAPTER 4

Talking pictures

Stuart Mealing

Koji Kabayashi (Kobayashi, 1986) looks forward to a time when the telephone system will automatically translate between the languages of the users in real-time. It will then be possible to speak in your native language to a Frenchman, a Japanese and a Russian and for each of them to hear your words spoken in their own language as you speak. More imaginatively, in *The Hitch-Hiker's Guide to the Galaxy* (Adams, 1979) it becomes possible to understand communication in any language by plugging a Babel fish into your ear, a fish with the natural ability to translate any received language into that of the wearer. Whilst these visions will not be realised for some years to come, the need for people with different tongues to communicate quickly and efficiently is increasing in inverse proportion to the speed at which the globe is being shrunk by trade and travel. New media used for communication, such as Multimedia and Virtual Reality (Mealing, 1992), carry with them implications of pan-global interaction which cannot be allowed to founder on the obstacle of exclusive natural languages.

Attempts to create international languages have broken down largely through the need for them to be learnt by a worthwhile number of people. There are, however, existing signs, symbols and icons which are understood internationally and a range of gestures, expressions and intonations which can intuitively convey a significant amount of meaning across natural language barriers. Indeed a picture, whether hand crafted or photographic, can contain a massive amount of information at a level of detail which spoken language rarely tries to match, although its interpretation might prove subject to local and cultural influences.

Images

Fig 1
The image aims to tell a story, but is open to interpretation away from the written article originally accompanying it.

The image above (Fig 1) aims to tell the same story as the article which it headed in *The Guardian* newspaper, namely that by drinking milk it is possible to avoid the crippling bone disease of osteoporosis, an affliction to which women are particularly prone. The centre box shows a crutch being abandoned as an arm (clad in a dressing gown sleeve of apparently female style) picks up the daily milk delivery from the front door. That central narrative is framed by two female figures, naked to suggest the "essential" woman. The style of the illustration is similar to a woodcut, and carries promise of unsophisticated honesty and basic good sense. Yet although this seems to me to be the image's message, it is obviously a personal interpretation largely developed from the written caption (Fig 2).

Well woman: osteoporosis

Such an image is very rich in possible interpretations when divorced from its caption, and carries with it nuances of meaning which are not readily translated into spoken language. It is this very richness which renders it unsuitable as a vehicle for the transmission of explicit information, although we can see that such an image could still be useful to us in cuing the domain in which a communication is taking place. A level of simplification and generalisation of visual imagery can, however, be achieved at which point the image can stand for a general case of thing and could be described as being both iconic and ontic. An icon is a symbol representing, or analogous to, the thing it represents (Collins, 1986), it is something which looks as though it could act as a proxy. Huggins (Huggins and Entwisle, 1974) asserts that: "Iconic communication deals mainly with non-verbal communication between human beings by the use of visual signs and representations (such as pictures) that stand for an idea by virtue of resemblance or analogy to it in contrast to symbolic communications where the meaning of a symbol is entirely nominal (such as English text describing a picture)".

Written language

It has been suggested that most current written languages, if not all written languages, are derived from pictorial sources. It is very easy to see how such a theory could be supported by looking at the examples below (Fig 3).

Figs 3
Examples
demonstrating how
written languages
may have been
derived from
pictorial sources.

43

雖如視

受話器置

あいうえお

▷ᴼ–ᴾⵁ◁ᴾ

आत्मीयता

    Clearly different levels of abstraction and stylisation develop, and at a high level of abstraction the language becomes composed of symbols, whose meanings must be learnt, or of individual characters which only acquire meaning when grouped together.

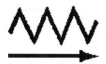 

Fig 4
Levels of
expressiveness

Adjustable resistor   Steam pressure   Slippery road

Icons

Whilst one can trace back to the pictorial origin of the *adjustable resistor* (Fig 4, left) image it is not as readily expressive as the *steam pressure* (Fig 4, centre) image, although the latter still requires the assistance of a label to offer up its meaning. The *slippery road* (Fig 4, right) image is so clear pictorially that the label merely confirms its meaning. In that order the levels of expression might be described as: symbolic, diagrammatic and illustrative. All three are likely to have the advantage of context to assist in their interpretation; the *slippery road* image, for instance, being seen within a red triangle, on a stripped pole, at the roadside, from a car, which limits its domain of reference comprehensively. We must not assume, however, that something we happen to understand has the same meaning or associations throughout the world, or that recognising the image necessarily explains how to respond to it.

The illustration above (Fig 5) shows four attempts to illustrate the Olympic wrestling event, presumably where the Olympic context was clear. Whilst the third image is the most simple graphically, and the most sophisticated visually, it is probably the least informative of the group. The first image, which is also the oldest, might seem naively pictorial but is the least ambiguous and hence the clearest. The efficiency of a close relationship between an iconic image and a picture is often neglected in the interest of graphical fashion or style, although severe simplification is often forced on the image by reproduction constraints (for example a 32x32 pixel matrix for a screen icon). Ideally the style should be timeless, but whilst this is unlikely to be possible, it is possible to produce images which retain the same basic content through style updates. Again, in this example, context and also familiarity are of vital importance. Whilst an Olympic wrestler would have little problem understanding any of the images, a stranger coming across them might well do so.

The styles of existing icons fall roughly into three main categories (Fig 6):

- silhouette (in front or side view), either solid shaded or in linear outline
- three-quarter top view, always linear and usually in isometric perspective
- 'realistic', drawn or semi-photographic (with graded shading etc.)

Fig 6
*The three main
categories of existing
icons.*

Whilst the latter category contains more information, that is not necessarily a virtue, as it detracts from its ability to describe a general case. In the above example, the left hand image can readily be seen to stand for *man* but the right hand image confuses that simple message by suggesting perhaps *business man, executive* or *young man*. It is also too loaded to be interpreted at a glance, and would therefore be unsuitable for use as a sign to be read from a speeding car. It might be useful to think of the presentation of its information by an image as having a temperature, from "hot" animated pictures through to "cool" static symbols.

Fig 7
*Examples of warm
(animated) and cool
(static) images.*

Cool presentation

Warm presentation

Fig 8
Icons requiring
familaiarity with the
subject.

It is easy to find confusing icons, sometimes requiring intellectual understanding (or a familiarity which might not be present) such as in the hare and tortoise (Fig 8), or open to misunderstanding such as in the toilet door example where the vertical line standing for a wall could be read as an indication that use of the facility is prohibited (Fig 9, left). The simple figure of a man has become widely adequate (Fig 9, centre) and does not require the relatively sophisticated knowledge needed to interpret the *toilet plan* icon (Fig 9, right).

Fig 9
The simple figure of
a man has become
widely adequate.

The power of universally recognised symbols can be harnessed in any dictionary of icons for as long as their meaning is perpetuated, even though their origins may no longer be remembered, as can other visual conventions that bridge language barriers. This obviously implies learning, and the point at which a set of icons starts to be comprehended will depend on the experience and sophistication of the user. Research will need to show the level of understanding that can be presumed in any given situation and the iconic language level for that context set accordingly.

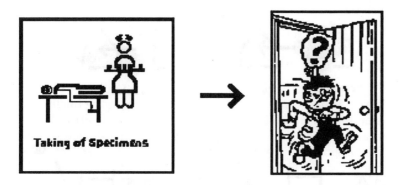

Fig 10
Stylised and comic
book images related
by an arrow.

The above image (Fig 10) shows, on the left, possibly the worst icon in the world, in which the image is stylised almost to destruction (and relies on its caption for interpretation). On the right a very "warm" comic book picture employs cartoon conventions to convey movement and panic, and utilises the widely understood *question-mark-in-a-bubble-over-the-head* convention. Between these two the potent arrow symbol is strong and concise enough to imply, in that context, a relationship between the activities in the images it divides. This example might also hint at our expectation about the layout of images on a page.

Visual languages

It is useful to look at some other attempts to create visual languages, even though they may be created for use in different contexts (such as for use with the handicapped). *Blissymbolics* (Fig 11) (Jones and Cregan, 1986) was first developed from ideas originating in the 1930s, combining representational pictographs, ideographic symbols, arbitrary symbols and international symbols and using positional layout and size to develop meaning. It requires a greater level of learning than we intend, but broaches interesting questions of syntax. How closely, for instance, should our iconic language mirror the construction of natural language?

Fig 11
Blissysymbolics.

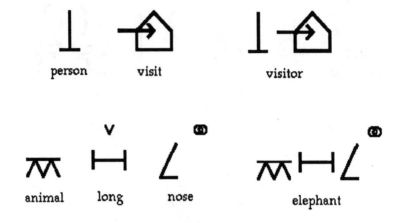

Worldsign (Fig 12) (Jones and Cregan, 1986) is interesting because it refers to dynamic actions and can itself be dynamic. I am sure that most people have discovered how much can be expressed in a foreign country with gestures alone, and also seen how expressive formal sign language can be. Also the layout of *Worldsign* is not tied to conventional linear rules. Both these features seem to be appropriate to development on the computer, which offers dimensions of presentation not available with other means. Whilst it is described as being able to be expressed in a number of ways, including computer graphics, I am not aware of it having been done so interactively.

Computer-based icons

Computers offer great flexibility in the presentation of icons, with size, position, colour and timing of appearance all easy to control. It is also practical to animate the contents of an icon or to animate the icon itself, and hierarchical icons are being developed (Mealing and Yazdani, 1990) which effectively offer a built-in help system. This allows the construction of more complex icons from basic icons since the compound icon can be interrogated to reveal its meaning (in terms of its evolution from the base images) - the icon knows its own history. It could also use still or animated imagery to explain itself. Whilst someone new to a group of icons would then have the security of being able to rely on their self-explanation, familiarity would render the exercise unnecessary. It is likely also, that the compound icons will be

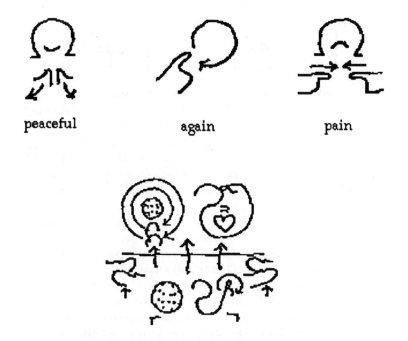

Fig 12
Worldsign.

peaceful   again   pain

Worldsign helps emerge from source
whole world interconnected feel-thinking

self-constructing and self organising, containing knowledge of rules governing their relationship to other icons. We can also extend the flexibility of meaning of icons by incorporating the principles of user modification and self modification.

In the first instance the language can be applied in a limited domain such as hotel booking, where the range of questions and responses is likely to be limited and predictable. A scenario such as this can be treated as having a 'situational script' (Schank and Abelson1977). It provides an appropriate starting point since language difficulties arise when even a limited range of ideas needs to be expressed.

Hotel booking

Hotel booking offers us the opportunity to apply iconic language in a simple dialogue - between potential guest and hotel. We recognise that in a working system it is preferable that the hotel computer should only offer *available* choices to the customer at each stage in the dialogue, but as this obviates our need to compile, send and reply to a message, we have assumed no interaction between hotel and customer during the compilation of the message and its reply. A typical scenario could find a stranger in a foreign city operating a touch screen in the window of a tourist office, or a traveller contacting a foreign town's accommodation bureau through his home computer terminal.

The compilation of the booking message is accomplished in stages and at each stage the current domain is cued by a picture resident in the background. In sequence these are: a 'typical' hotel front (Fig 13), a 'typical' hotel reception area (Fig 14), and a 'typical' hotel bedroom (Fig 15), each new screen holding the background picture for a second before the information is faded in over it . Therefore when dealing with the required room type(s) the background picture on screen would be of a room. Sub-domains might later require a typical bathroom, dining area etc. and in the future such images could even be live vide.

The first screen (Fig 13) shows a hotel overlaid by an appropriate caption, and *clicking* anywhere on the image starts the booking sequence. (This prototype is created in HyperCard™ and is operated with a standard mouse, but conversion to other input devices, such as a touch screen, is anticipated by the design.) The screen then invites input of destination, cued by a map, (currently typed in, but could be from selection of map locations) and selection of hotel type, by selecting from cyclable 'star' ratings. Movement to the next screen is initiated by *clicking* on the 'tick' icon, a convention observed throughout the package.

The second screen (Fig 14) shows a hotel reception area and invites selection of the dates and times of arrival and departure. (As it stands, the days and months are expressed in natural language which does not satisfy our ends. A numerical layout was avoided as there is a conflict between the order in which days and months are laid out in different countries, and as it was important to us to reach the point at which we could send a message: this is a problem to which we will return.) The number of nights that have thus been booked is indicated by black bars

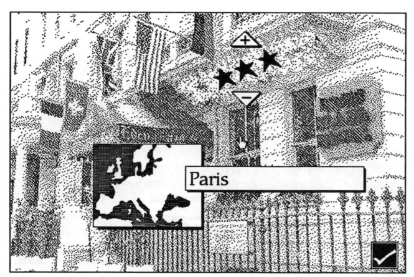

Fig 13
A typical hotel front.

which appear (and disappear) as each night is added (or removed), but whilst this added display is useful its meaning is not immediately clear in its present form. The 'tick' icon moves the user to the next screen.

The third screen (Fig 15) shows a room overlaid with icons permitting the selection of room type(s). Four icons each 'unlock' further related icons to enable (a) the number and type of occupant to be shown (b) the number and type of beds required (c) the type of bathroom facilities required and (d) a range of other available facilities (such as TV) which can be selected. Throughout the application, the user is presented with a limited range of choices at any one time. The features are selected by *clicking* on the relevant icon, which produces a clone beside it, and then *dragging* that clone into the room which is shown as a rectangle, these mobile icons can also be deleted. More rooms can be requested by *clicking* on the '+' and '-' icons (up to a maximum of 4 rooms in the prototype) and the rooms' occupants and contents can be rearranged to suit. This method of presentation was felt to allow the user more flexibility in organisation than if each room had to be defined separately, and whilst no controlled tests have yet been conducted, the interface has been found to be intuitively obvious. In order to cue the user's understanding of the principles of organising a room, the screen opens showing one room with occupants. When a satisfactory arrange-

Fig 14
A typical hotel
reception area.

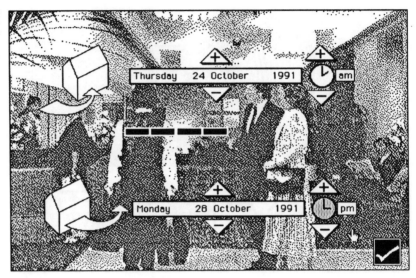

Fig 15
A typical hotel
bedroom.

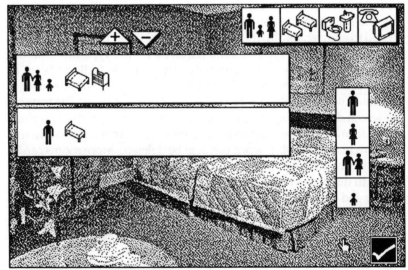

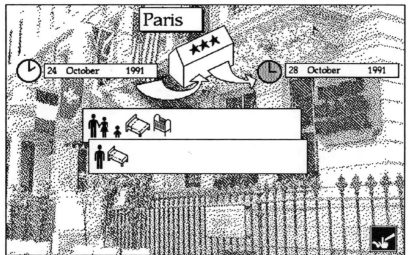

Fig 16
The complete
booking
requirement.

ment of rooms, occupants and facilities has been achieved, the 'tick' item moves the user to screen four (Fig 16) which displays the complete booking requirement. If this is satisfactory, a further 'tick' sends the message to the hotel.

The message is revealed to the hotel in stages (Fig 17). Confirmation of the acceptability of each part of the message (by selecting a 'tick') moves on to the next part of the message, unavailability (indicated by selecting a 'cross') brings up the range of available alternatives. In Fig 18 the choice of 'star' rating was unavailable and the alternatives are presented so that an alternative can be offered.

In Fig 19 it can be seen that the chosen bedding arrangement for the first room is not available, and two single beds are offered instead of the double bed requested. The large ticks and crosses indicate to the customer what is, and is not, available and the question mark precedes the alternative which is being offered ("Is this OK?"). The layout of the rooms and alternatives is intended to enhance their understandability. The cursor is about to select a 'tick' to indicate that the second room is available as requested.

The final message (Fig 20) is sent back to the customer who will then be able to accept or reject the alternatives offered, continue the dialogue, and confirm a booking. The application does not pretend to be comprehensive or the most practical in real terms, but is an initial attempt

Fig 17
*Screen showing
required dates of
stay.*

Paris   24 October 1991 ⇨ 28 October 1991

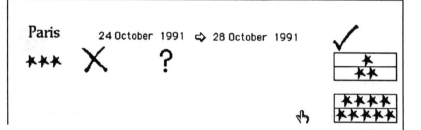

Fig 18
*Screen showing
unavailable and
available 'star'
ratings.*

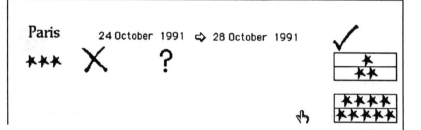

Fig 19
*Screen showing
available bedding.*

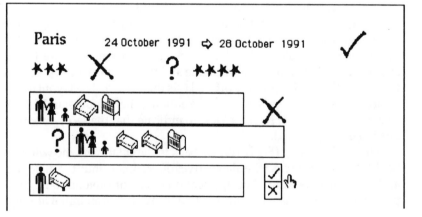

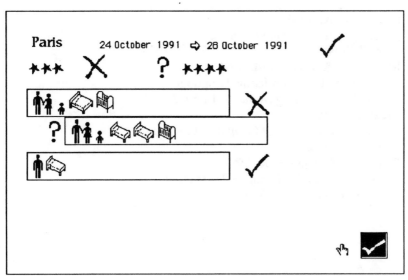

Fig 20
The reply which is
sent back to the
customer.

to create a simple, interactive, iconic dialogue using hotel booking as a theme. It does, however, offer much that could be used in a real system, and serves its purpose in starting to explore the possibility of communicating with icons.

References

Adams D. 1979
    The Hitch-hiker's guide to the Galaxy
    Pan Books Ltd.
Collins English Dictionary, The 1986
Huggins W. H. & Entwisle D. R. 1974
    Iconic Communication: an annotated biography
    The John Hopkins University Press
Jones P. R. & Cregan A. 1986
    Sign & Symbol Communication for Mentally Handicapped People
    Croom Helm

Kobayashi K. 1986
Computers and Communications
MIT Press

Mealing S. & Yazdani M. 1990
A Computer-based Iconic Language
Intelligent Tutoring Media Vol 1 No 3

Mealing S. 1992
The Art and Science of Computer Animation
Intellect Books

Schank R. & Abelson R. 1977
Scripts, Plans, Goals and Understanding
Lawrence Erlbaum Assoc.

CHAPTER 5

Generalising language tutoring systems

Masoud Yazdani and Jo Uren

Traditional Computer Assisted Language Learning (CALL) systems are made out of masses of pre-stored patterns of potential interactions, therefore they cannot recognise or comment upon errors encountered, even if the errors are frequent, unless these have been individually and specifically anticipated by the programmer. They are not readily modifiable by teachers either; as the flow of control (programming information) is not separated from the knowledge of the domain.

Instead, we have been trying to produce systems which are capable of being used by teachers in a variety of ways. Our aim is to build tools for learners and teachers; providing hybrid systems made out of human teachers and computational knowledge-based systems.

The roots of our endeavour can be found in FROG (Imlah and du Boulay, 1985). FROG had a general linguistic knowledge of the vocabulary and the grammar of French and therefore could deal with a wide variety of unanticipated input from the user. The main problem we found with this system was that the knowledge and the routines (the control) for processing the input were so intertwined that no one without an intimate knowledge of FROG could extend its functionality.

We built FGA (Barchan *et al.*, 1986) in order to address some shortcomings of FROG. FGA, attempted to keep the knowledge of the grammar, dictionary and the error reporting routines as distinguishable data structures separate from the processes needed to deal with them and the users' input. Although FGA had some successes in the demonstration mode, it was not used by any practising teachers as basis of a course of tuition in French.

Our attempts at making FGA more general, and easier to use by people without any programming experience lead to a new system called LINGER (Language INdependent Grammatical Error Reporter) developed by Barchan (1987). In the process, we feel we have made LINGER general enough to be a good basis for other Romance languages.

Since 1987 LINGER has been used as a prototype teaching tool at Exeter University's French Department and has been the subject of a series of interrelated projects to evaluate and extend its potential as a basis for the teaching of Modern Languages by others not directly associated with the research team.

The architecture of LINGER needs to incorporate the following components:

- a well structured linguistic grammar
- knowledge of deviations (exhibited by novices) from the "correct" grammatical structures and the associated remedial advice.
- a flexible dictionary.

LINGER when supplied with these sources of knowledge reports and corrects any grammatical errors encountered in the user's input. The intention is that LINGER should be easily configured for a particular language by an "expert" in that language but not in computer science. That is, it should serve as a tool for the language teacher.

LINGER is fulfilling its language independent objective; it supports French and is currently being extended for Italian, Spanish, German and English.

Given the high modular approach adopted by LINGER how can a non-computer scientist do anything with it? Could language independence still be maintained and LINGER still be capable of efficient parsing of inputs and reporting the errors? We shall present some tentative indications after we have shared our experience of trying to extend LINGER for Spanish. A few examples will be given to show how LINGER has coped with this language.

In the following sections we shall present the Independent Language databases for Spanish as can be seen from Fig 1.

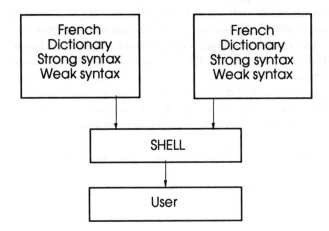

Fig 1
LINGER Structure.

We plan to illustrate the basic principles underlying the grammar dictionary modules of the system. Examples will be used to show how a grammar writer may approach the task of writing his own modules. Each section will be divided into a brief discussion of the module, a definition of LINGER's specification and an example of the approach adopted for the building of the Spanish modules.

System architecture

Four distinct design features typify the goals and strengths of LINGER. These are as follows:

1. Modularity & extendibility as dictated by the requirement of language independence.

2. Generation of correct sentences as desired both in a teaching environment and in the long term extendibility of the system.

3. Handling of unknown/incorrect words.

4. Grammar writer's control over issuing and content of error messages for pedagogical purposes.

A configuration of the system consists of three main modules: the language specific dictionary, the language specific grammar (strong/weak syntax) and the language independent shell. Each module will be briefly described. However, for further implementation details see (Barchan, 1987).

The distinction of dictionary, grammar and parsing mechanism is a virtual requirement of a language independent system. Central to LINGER therefore lies the two flexible formalisms of:

1. language dependent grammar and dictionary

2. language independent shell

which when combined provide a powerful tool.

We shall now move on to consider the three main LINGER modules in detail.

The dictionary

The dictionary is one of the two language-dependent data files required by the language independent shell to function for a particular language. It serves two purposes: firstly, and most obviously from its name, it contains all the words in the specific language which are known to the system; secondly, it holds information relevant to each word including what modifications can be made to that word together with their significance. This information is usually, but not necessarily, grammatical in nature, but there is a distinction between this information and that contained in the language's grammar file. In the latter, the information is concerned with what legal grammatical structures (e.g. sentences, noun-phrases, verb-phrases etc.) may be formed in the language and how they are put together, while in the former the concern is with what individual words are permissible in the language, how they may be modified and what significance each such modification entails. Hence the distinction is that the grammar file specifies the language's non-terminals, while the dictionary file deals with the precise form which the terminals may take.

The grammar

The grammar file is the second of the two language-dependent files required by the language independent shell to function for a particular language and includes both the strong and weak syntax. It serves two functions: firstly, to permit the grammar writer to specify what grammatical constructs exist in a given language and how they may be combined to form legal sentences, noun phrases, verb phrases and so on in the language. We call this 'strong' syntax. Secondly, it allows him to indicate what rules must be obeyed to produce correctly formed sentences within the general framework of the constructs permitted, such as appropriate numbers, gender and so on. We call this 'weak syntax'.

If the grammar writer wishes to anticipate certain common errors he may include messages to be presented to the user if the input exhibits the appropriate features. This facility normally referred to as the "bug catalogue" is embedded in the weak syntax. As yet LINGER does not allow the creation of an independent bug catalogue knowledge base.

The language-independent shell

The shell is the language-independent core of the system. The shell contains routines for such actions as accepting the user's input, consulting the dictionary, interfacing with the grammar, attempting to parse the input, reconstructing the sentence correctly, comparing the new version with the input sentence and producing its final judgement to the user. The behaviour of the shell can be viewed as consisting of three stages: pre-parsing (analysing the words), parsing and choosing the 'correct' sentence.

Although the shell's remedial capability is dependent on the coded knowledge of common mistakest his knowledge is coded independently from the programming details. In addition the shell enjoys the benefit of some language- independent "heuristics". These are the common ways in which people make mistakes in any language leading to their sentences becoming more like a word soup. The current word soup heuristics include:

- word in wrong place
- word missing
- word of Wrong Type
- word shouldn't be present
- one word shouldn't be present and another word missing.

Generalising LINGER

LINGER is based on a Definite Clause Grammar (DCG) notation (Pereira & Warren, 1980) and is implemented in Prolog. DCG is simple to understand and easy to modify. One of its features DCG is that basic principles are defined before more complex concepts are considered.

The grammar is divided into two parts: a "strong" and a "weak" syntax. Also referred to as a "grammar specification" and "checks". Grammatical categories and specific attributes which may appear within the grammar rules, checks, or even the dictionary are essentially arbitrary tags selected by the writer of the files to encode the characteristics and behaviour of the language.

Strong syntax

To describe how a grammar can be expressed, the following notation is used. Each rule has the form:

nt --> body

where <nt> is a non-terminal symbol and body is a sequence of one or more items separated by commas. Each of them is either a non-terminal symbol or a sequence of terminal symbols. The meaning of the rule is that 'body' is a possible form for a phrase of type 'nt'. As in the syntax of clauses, it is possible to allow this basic notation to be extended by allowing alternatives to appear in the 'body'.

In order to understand how the DCG representation serves as a basis for a straightforward top-down parser it can be best viewed as a "sentence building" process which consists of the repeated decom position of a <lhs> into a <rhs> until appropriate terminals are extracted to satisfy rules.

The syntax for the French grammar is given as:

    <lhs> --> <rhs>
    where <lhs> is <name>(formed(<name>,[<variables>])),
    <name> is the name of the non-terminal,
    and <variable> is a list of name.

Each one will be used in the same order in the <rhs>,

    where <rhs> is either [] or <name1> (<var1>), <name2> (<var2>),
    <name1>, <name2> are the names of the <lhs>.
    and <var1> <var2> are the variables in the <lhs>.

To take a simple example the syntactic representation:

    noun-phr(formed(noun-phr,[D, A1, N, A2])) -->
        determiner(D),
        adj-list(A1),
        noun(N),
        adj-list(A2).

would yield sentences of the French grammatical form:

This form of representation is followed as the basis for the building of the grammatical representation of the Spanish language.

To construct the required interpretation for Spanish an example will serve to illustrate how a grammar writer may set about the task. Given the Spanish sentence: la chica es guapa (The girl is pretty). This can be broken down into the following constituents:

These rules taken from the tree can then be translated into the grammar as:

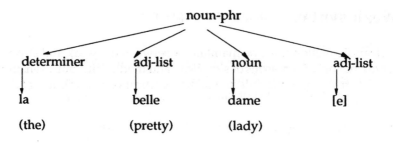

```
sentence(formed(sentence, [VP])) --> vp(VP).
vp(formed(vp, [NP, V, A])) --> np(NP), v(V), adj(A).
np(formed(np, [D, N])) --> determiner(D), noun(N).
noun(formed(noun,[N,Pos])) --> it_is(noun,N,Pos)).
determiner(formed(determiner,[D,Pos])) -->
     it_is(determiner,D,Pos)).
```

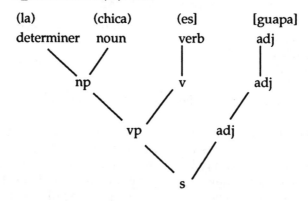

```
verb(formed(verb,[V,Pos])) --> it_is(verb,V,Pos)).
adjective(formed(adjective, [A, Pos])) -->
     it_is(adjective,A,Pos)).
```

Having considered how a grammar writer may deal with a strong syntactic rule specification it remains necessary for the writer to enter the weak syntactic rules which are viewed as constraints or checks upon the language.

Weak syntax

A clear distinction must be maintained between the weak and strong syntax not only for reasons of simplicity and modularity but to ensure that the strong syntax is reflected in the specification of attributes of the weak syntax. That is, given a strong syntactic rule of the form:

sentence --> subject, verb

if the attributes of the language are such that the declarative reading is 'for a sentence to be correct the main verb must agree with the subject' the grammar-writer can impose those features of the grammaticality of the language by adding 'checks' into the weak syntax module.

The specification of the weak syntax is essentially that of:

check([<structure>], <scope>, [<precondition>], <requirement>).
   where:
[<structure>] is a list of the structures to which the requirement is to
   be applied.
<structure> is either name of <lhs> of the strong grammar rule or
   the name of <lhs> of the strong syntax.
<scope> is the <structure> within which the check is to be
   performed.
[<precondition>] is a list of <conditions>'s which must be true
   before the requirement is performed.
<requirement> is the action to be taken by the check.

Each check needs to provide the following information:

i.   The grammatical type of the terminal which is to be checked.

ii. The sub-structure within the sentence across which the check is
     to be made.

iii. Any preconditions which must be met if the check is to be valid.

iv. The name of the check to be performed, together with relevant
     parameters.

When the weak syntax specification is applied to the case of adjective and noun agreement in Spanish, the rule given below, ensures that within a noun-phrase the determiner and adjective should agree in number and gender with the noun. It can be read as "check that the determiner and adjective in a noun-phrase agree in number and plurality with the noun."

```
check([determiner, adjective],
noun-phrase, [],
concord([gender,plurality],noun,[])).
```

There are further syntax specifications for <condition> and <requirement> which are not discussed here. Interested readers should consult Barchan (1987).

Lexical processing

There are no problems in the adaptation of the French dictionary and categorisation of words to handle the Spanish language. However, related problems are foreseen where the same word may be classified in one or more ways, for example in instances when a word with membership of one class (i.e. noun) can also belong to another class (i.e. adjective).

A decision also needs to be made regarding the scope of the dictionary. Should it be of a "limited domain" or "unrestricted"? The former would concentrate on a particular text limited to a topic whilst the latter would involve the insertion of words most frequently used in the language.

At the present time the system can cope with most words not in its dictionary and assign them to grammatical classes with a high success rate. A possible useful extension would be to store unknown words encountered, so a system administrator (i.e. a teacher) can add them later to the dictionary.

As any dictionary writer will confirm, there is a need to distinguish between words of the language and the syntactic classes to which they belong such as noun, verb, etc. The dictionary configuration for LINGER is considered to be simplistic yet powerful. For reasons of optimal generality there are three distinct kinds of entry to be found in the dictionary:

1. information about grammatical categories

2. individual words

3. sets of general endings for various word-types in the language.

The syntax specified for the dictionary falls into 3 specifications: Grammatical/Word Endings ; Word Specification (including word root, grammatical type); and Word Ending Types. These enable words to be entered readily. For example, given the attributes of a Spanish noun as having 'gender', 'number' and 'person', the entry for the Spanish word 'barco' (boat) would be:

```
info(noun, [[plurality(s)], [plurality(p)]], [gender],
[person(3)] ).
giving rise to the further entry:
word("", ["barco"], noun, ["","s"], [m] ).
```

It is possible to generalise these patterns of word-ending types. However, we can enhance the specification for this class by adding:

```
info(gp_end,[[gender(m),plurality(s)],[gender(m),plurality(p)],
[gender(f),plurality(s)],[gender(f),plurality(p)]],[],[]).
ending_type(gp_end,["o","os","a","as"],[]).
```

which states that the group endings for the class of nouns can be masculine or feminine, singular or plural and that the ending types for each of the Spanish attributes are 'o' (masculine, singular), 'os' (masculine, plural), 'a' (feminine, singular), and 'as' (feminine, plural).

At the lowest level little attention has been paid to morphological considerations. It should be possible to incorporate a morphological analyser in the module at a later stage.

Semantics

LINGER currently ignores any semantic considerations. Therefore it would fall victim of sentences such as "Colourless green ideas sleep furiously". If language is concerned with communication and grammar is concerned with "correctness" of "form" then we need to incorporate LINGER within a wider educational environment. Some of these "environmental" issues are discussed further in Yazdani (1987), while others are being studied in more detail by Uren (1988).

Conclusion

The questions posed in this attempt at generalising LINGER from French to Spanish indicate the extent of the work which lies ahead and the number of problems still to be addressed. For the moment we plan to build on the limited success of LINGER as a prototype which will be tested comprehensively within the next two years. LINGER, however, is a workable system fulfilling its early objectives in its present form:

1. It has a language-independent shell and language subsets for French, German, Italian and Spanish. Work is developing (O'Brien and Yazdani, 1988) for an English system.

2. It is robust and adabtable enough for it to be the core of future work.

However, it is limited in its capabilities and 'intelligence. Further developments needed include:

i.   modularity and extendibility of linguistic knowledge

ii. multiple error handling

iii. separation of weak/strong syntactic errors

iv. treatment of unknown words

v.   treatment of incorrect endings/ spelling errors

vi. improvements to the parsing techniques.

There are a number of extra modules which can be added to LINGER in order for it to become a fully practical proposition for the teaching of language. New ideas to be considered include:

1. Student Modelling: adapting the systems behaviour to the individual user.

2. Explanation Module: why learners make the mistakes they make; how best to explain the errors to them.

3. Teaching Strategy Module: how to plan a course of lessons; how to use LINGER as a teaching tool; how to build the educational environment around LINGER.

4. Machine Learning: how to build a catalogue of bugs automatically.

5. Contrastive Mother tongue Module: how to use the knowledge of the user's mother tongue in more effective error finding, explanation and lesson planning.

Our progress towards LINGER can be viewed as a move from constructing tutoring systems for specific tasks towards creating tools of an even greater generality. LINGER has been designed to be as general as possible, making few assumptions about the nature of language for which it might be used. It is expected that future identification of areas of similarity between different languages may lead to the development of more systems which can use LINGER as their core but augment it with assumptions about a language or particular methods of teaching that language.

Acknowledgments

The work reported here is sponsored by a grant from the Economic and Social Research Council (ESRC). We are grateful to our colleagues Paul O'Brien, Keith Cameron and Judith Wusteman for their continuous support.

References

Barchan, J. 1987
Language Independent Grammatical Error Reporter
M.Phil. thesis, University of Exeter

Barchan, J., Woodmansee, B.J. and Yazdani, M. 1986
A PROLOG-based Tool for French Grammar Analysis
Instructional Science Vol. 14 pp 21-48

Chomsky, N. 1956
Syntactic Structures
Mouton, The Hague

Imlah, W. and du Boulay, B. 1985
Robust Natural Language Parsing in
Computer Assisted Language Instruction *System* Vol. 13 pp 137-147

O'Brien, P. and Yazdani, M. 1988
eL: A Prototype Tutoring System for English Grammar (Error
Detector and Corrector), the *Proceedings of the Third International
Symposium on Computer And Information Sciences*. New York,
NOVA Science Publishers, Inc

Pereira, F. & Warren, D. 1980
Definite Clause Grammar for Language Analysis
Artificial Intelligence, Vol. 13 pp 231-278

Uren, J. 1988
Teaching Strategies: Languages and LINGER
Research Report, University of Exeter

Yazdani, M. 1987
Artificial Intelligence for Tutoring
Tutoring and Monitoring Facilities for European Open Learning
J. Whiting & D.A. Bell (eds).
Elsevier Science Publishers

CHAPTER 6

Dual purpose learning environments

Robert W. Lawler

Feurzeig (1987) described his view of "intelligent microworlds" as permitting the mode of interaction between the user and system to be switched (by the user) from exploratory to tutorial to evaluative. The view offered here derives from Feurzeig's suggestion but moves in a different direction, to focus more on the purposes of the parties involved - instructor and student - than on the performance mode of the system. Consider the following as an example of a system that will permit dual usage.

When engaging in explanation of grammar, a teacher wants to offer his students a lucid and well-articulated description of the principles which govern the forms of a language in use, along with succinct examples illustrating those principles. If an intelligent system supports such use, one can have instructional use of the system. Students often prefer a more exploratory approach to learning. One might say, for example, "Let me try to do something novel to find out what I can do with this language" or "Let me probe what the system can do, beyond requiring me to generate a sentence it will accept. Can I determine what the limitations of the system are ? Can I improve the system's grammar in such a way that it will be more nearly perfect ?" Pursuing such questions puts a student in a very active mode, one in which some students will learn much better than most other ways regardless of the domain or language focus.

A system with such flexibility in use would be a dual purpose learning environment, one in which the instructor can have his way, provide his best guidance, but hopefully one in which the student can also act in a powerful and positive way to correct and augment the system itself and through doing so develop his own knowledge as much as he cares to.

Project objectives

Observing that people in Europe have been more sensitive than Americans to the need for learning multiple languages and to second language instruction, we tried to take advantage of their expertise. A primary objective of this project was to port a European PC based instructional system, LINGER, for use on Macintosh computers in the USA. LINGER (Yazdani, 1990) is an application package directed to instruction in several foreign languages. It is a Prolog-based intelligent tutoring system. The original version of LINGER was a research vehicle. We tried to scale up that system for wider use. Adapting LINGER for the Macintosh had two dimensions. First was the issue of the Macintosh port. We chose LPA Prolog as the implementation language, expecting little trouble in converting the source code. So we found the case. Some low-level routines, primarily based on I/O calls had to be re-written. Interface improvements, especially when set up to take advantage of Macintosh features, required additional coding.

The second dimension of the project revolved around scaling up the size of the dictionaries and grammars used with LINGER. Although this sort of activity is commonly undervalued, it often reveals problems not apparent with small-scale prototypes. LINGER has three main components: a dictionary representing the words of the language, a database of rules which amount to the grammar specified for that language, and an inference engine which uses the dictionary and the rules to test the grammaticality of strings of words submitted to the program.

LINGER as an ITS

LINGER was initially designed as a grammar checker for novices at a second language. Artificial Intelligence (AI) techniques were needed in LINGER because novice text production typically deviates considerably from the standard of the target language. We use ITS (for "Intelligent Tutoring System") here in a loose sense to mean that LINGER uses AI programming technology for instructional purposes. In effect, LINGER was designed to guess the user's intended text from the input then judge the closeness of fit of the entered text to a correct expression of the inferred intended text. LINGER is in character more analytic than didactic.

When LINGER receives a string of text, it returns to the user a parse of the string with comments and suggestions about the string's validity. It also attempts to compose its own version of what the string should have been had the user produced a grammatically correct version. This may seem an audacious goal - unless one considers the limitations of the original context of use. LINGER was created originally for a foreign language instructor who was tired of correcting novice-students' obvious errors. He hoped that his students could improve their assigned essays by typing sentences into LINGER and receiving grammar criticism at a fairly low level of sophistication. LINGER was intended to be used as a kind of "language calculator" to catch obvious errors. It was necessary for the system to deal with unrecognized words because one could not count on students typing correctly.

Grammar and LINGER's architecture

The prototype grammar of LINGER is a set of Prolog rules. The prototype grammars of the three ported LINGER systems (Spanish, English, and French) are largely similar within the limited domain of the grammars' coverage of the languages. The prototype grammars were intended as an examples to be changed and developed by the final user. Even given such an intention, one needs describe the starting point to see what progress is possible. The character of the grammar can be judged from its depth, size, breadth, and modifiability. Consider the English grammar as typical of the other LINGER prototypes. It is five levels deep. The English grammar has four levels of structural rules (a fifth set, "checks", verify grammaticality after structure has been determined). There are sentence-to-clause rules (2 in number); clause-to-phrase rules (4); phrase-to-phrase rules (23); and word-lookup rules by part of speech (17). The number of checks is 28. The size of the grammar, in total number of rules then, is 74.

Simple dictionaries require complex processing rules, and vice versa. Decisions about grammar rules and their coding interact with the knowledge representation used in the dictionary. In LINGER, for example, non-standard plural inflections are coded directly into the dictionary. So also are bound comparative forms of adjectives. This representation decision has two primary consequences. First, multiple dictionary entries are needed for those words which can serve as different parts of speech. Second, one should expect a persistent trade-off in the implementation between extending the dictionary (and thus parsing time of entered strings) and extended parsing times through increased complexity of processing.

The issue of breadth is complex, in that it measures the extent to which the rule set covers the grammar of the language. The LINGER grammar prototypes are narrow in some ways which can be easily modified and in other ways which require a major redesign of the system. Consider the easily modifiable cases first. In English, the passive is formed with a past participle and an auxiliary verb. LINGER's prototype dictionary recognised only forms of the verb "be" as a passive-forming auxiliary. English permits the use of two other auxiliaries, "get" and "become" to form either action-oriented or developmentally focused passives; for example, "the thief got caught and in due time and with due process became imprisoned". Such omissions can be simply corrected at need by adding definitions of the two auxiliaries to the dictionary. The more complex limitations of LINGER derive from interactions of the dictionary, grammar, and the inference engine.

Macintosh LINGER dictionaries

At the conclusion of this project, dictionaries available for use with Macintosh LINGER in three languages have been scaled up by a factor of 20. Original LINGER prototype dictionaries were a mix of parts of speech totaling between 50 and 70 words. The new LINGER English dictionary is approximately 1200 words and is implemented as three separately loadable Prolog files, of approximately 200, 500, and 500 words each. The new LINGER Spanish dictionary is 1300 words long. It is implemented as eight separately loadable files. The French dictionary, of approximately 1000 words, is implemented as a single large file. It is a word collection typical of those used for vocabulary review in US high school French courses. It is not partitioned because the source vocabulary did not include any principles justifying grouping of the words in a usage-based, meaningful way.

The Spanish dictionary is based on Keniston's (1920) collection of *Common Words in Spanish*. The list derives from dialogue appearing within plays and novels. It was an early attempt at representing real conversational vocabulary and a reasonable one given the absence of recording equipment. The words are grouped in eight levels of use by frequency. The source frequency partition explains why one could break-up that list into groups which might be of a manageable size for study and/or instruction. One reason for choosing that list was that for more than 50 years the Keniston word list has been used by text book publishers of commercially available textbooks in Spanish; in this specific sense it is still "state of the art". Some more recent frequency counts have been based on interviewing people (this has a natural appeal as far as realism goes), but they are not vastly different from the Keniston list.

The English dictionary is derived from frequency counts of words appearing in a collection of stories told by children. It is based upon a study of children's' story telling by Moe *et al.* (1982). They asked children to tell their favorite stories, then counted the frequency of occurrences of all the words. We took that study as a basic source, then deleted references to individuals (Spiderman, Superman, Goldilocks, etc.) to create our own list of 1200 frequently used words. Since the words were produced by young children, this list might be especially suitable as a list of very common words in English. The original collection of words has been been somewhat compressed by removal of duplications based on inflections and contractions.

The use of polylingual ITS with other media

How should we think of a long-term sequence of language learning activities in which such interactive teaching systems and tools could play a productive role ? We need a practical view with at least one component that takes full advantage of the technology and yet also respects the limitations of expense and cost that technology involves. How could future LINGER-like systems fit in with less expensive, more traditional educational technologies and practice?

Let's suppose a language training program involves an immersion experience at a hypermedia-capable language training centre. Such could be a place for total immersion in the second language, where people would speak and listen to the second language as well as working with systems for second-language instruction. Before people attend such a centre, it would be important for them to be familiar with the kinds of systems they would work with, what such systems could do and what such a systems' goals were. It would be efficient if future centre students could be introduced to the training system through remote site viewing of videotapes about them. The optimal way to do so would be to strip from the interactive system introductory demonstrations made in the native language of the future student, so that, when at the training centre, the student could concentrate on use of the system with the target language of instruction.

When people leave the training centre and return to their normal positions, they might then find a LINGER-like facility useful primarily in the mode of a linguistic calculator. A LINGER-like system with a nearly complete grammar - designed for maximum processing efficiency and NOT using any hypermedia training materials - would be most cost effective. At this time, an interactive teaching system fits a niche within a larger language training and learning program.

LINGER usability for future instruction

The LINGER system we have discussed is not usable in classrooms today, but it serves as an interesting and promising research tool. What sorts of utility might it have after a significant redesign and further development ? The target audience for LINGER is one of people who are learning a second language. One may think grammar is important as a crutch for learning second languages - not necessarily because of the good fit of grammatical rules to what is in the mind but primarily because such systems of rules have been of proven value to people in making judgments about how writing and talking should proceed. Grammar will continue to be a subject of instruction while second languages are taught.

One strength of LINGER is its ability to continue processing even when it encounters words not encoded in its dictionary, using the structural rules to guess at the type of unknown words. Such flexibility is essential for educational applications. A major new objective for future LINGER systems should be extension of this capability to permit the addition of user-defined rules to the grammar. Adding such a capability will not be easy, but it should be possible through fusion of techniques based on programming by example and through definition of erroneous variations of new structures as "near misses" at the time of grammar rule extension. There are three broad categories of application we can foresee now for such systems as LINGER or its descendants:

- traditional instruction
- student guided discovery
- instructor experimentation.

Traditional instruction

One can imagine LINGER systems as linguistic calculators, using which a student might verify that his composed sentences are correct before commiting them to paper in an essay. Such could increase the feasibility of writing assignments in second languages (and thus enjoy-abiliy for both student and instructor). By itself, this would be a significant enhancement for many language-instruction programs.

Student-guided discovery

One of the key questions in education is the extent to which students are actually active in learning what they are studying. We believe this approach to the use of educational technology holds the most promise for individual students through engaging them actively in their studies, but making this approach effective will require redesign of LINGER systems. In self-guided discovery mode, the student would use a LINGER-like system as a modifiable grammar, one whose operations and performance he could explore and change.

Instructor experimentation

Experienced instructors try to diagnose the errors their students make so that they may offer them effective advice on how to improve their grasp of the language. One approach in current use is to develop lists of common errors and offer them to students as warnings of what they should avoid. Lists such as that of Tables 1 and 2 might serve future ITS as bug catalogues. LINGER could enhance an instructor's ability at diagnosis in a way that goes beyond lists of errors or bug catalogues by providing a developing language modelling capability. This could enhance the development of teachers' intuition and their explicit knowledge of the roots of errors manifest in speech production.

Table 1
Categories of
Common
Structural
Novice Errors in
Spanish [1]

Concordance	Passive
- gender	- true
- subject-verb	- Se passive
- noun-adjective	Negatives
Word Order	Apocopation
Redundancy	Nominalization
- subject pronoun	- adjectives
- indirect object	- possessives
Pronouns	- infinitives
- direct	Substitutions & Confusions
- indirect	- ser / estar / haber
- reflexive	- por / para
- demonstrative	Personal a
- relative	Prepositions
Subjunctive	- bound
- present; past	- simple
- impersonal	- complex
- noun clause	Noun Markers
- adverbials	- definite
- adjectives	- indefinite
- conjecture	- otro
Imperatives	Possessives
- participles	Numerals
- present	Adverbials
- past	Conjunction
Tense/Morphology	Augmentatives/Diminutives
- present	Hace + time
- preterit/imperfect	Comparatives
- perfect	Más que
- future	Tanto como
- conditional	Infinitives
- progressive	- verbal phrases
Interrogatives	- prepositional
- Qué/Cuál	

[1] Unpublished. Developed by Professor Flint Smith, University of Syracuse, USA.

Even now, in LINGER's early state of development, one can load with the shell, the grammar of one language with the dictionary of another, as the dictionaries and grammars are modules. One ask may ask then what sort of performance would come out a system with a "native" grammar of one language and the beginnings of a vocabulary in another language ? What would be the results, in terms of the performance of the system, if one began to add to the native language a rule representing a specific grammatical construct of the grammar in the second language ?

But how does that relate to what students and teachers actually do and learn ? "*Me llamo es," says one of the thousands of beginner Spanish students that pass through one's class–and it appears that there are two approaches to understanding this common pattern (with an eye towards its easy eradication). We can use a list of commonly found incorrect patterns, such as that of Table I, to locate an explanation of precisely what the error is then correct the student with a lecture on the need to use a reflexive verb construction to identify oneself in Spanish. Any experience as a language teacher will convince you that this is almost useless. Clearly, some other approach is needed and perhaps a better understanding of the nature of the error would help.

One road to that understanding is to be seen in the three kinds of analyses identified by James (1990) as "learner language". These are contrastive analyses (comparisons of native language to target language), error analyses, (comparisons of what Selinker (1980) called interlanguage to a specific target language), and transfer analyses (comparisons of interlanguage to the learner's native language made in an effort to find evidence of inappropriate transfers from the native language to the interlanguage). Certain of these analyses have had important effects on language teaching practices. For example, contrastive analyses have, in the past 30 years, had their effects on the ways some language teachers generalise about such items as aspect of verb tense in Romance Languages; and error analyses have revolutionised the ways some language teachers respond to error. There may be the possibility for using LINGER-like systems to effect another analytical approach which may have a similarly beneficial effect. LINGER now accommodates more kinds of grammatical structures in more languages that it used to. Further development will lead to improvement and, perhaps, a new outlook on the analyses mentioned above.

Table 2 *Expansion of A* *Single* *Concordance* *Novice Error* *Description* *(Spanish)* [2]	Nouns listed in the knowledge domain as masculine or feminine require an accompanying determiner and adjective of the same gender.

la Mañana = the morning. This is the way to say "the morning." Changing the determiner to el usually makes no change in meaning. It only creates something that doesn't exist in standard Spanish, so we call it an error. (Here, the change would be meaningful. El mañana can be used to say "the future" in a general sense).

* Es una mañana hermos(o)s vs. Es una mañana hermosa.

A common error is illustrated above. Using the wrong ending on the adjective does not provide for concordance of gender.

Here, the example shows the addition of tres (three)

* ...tres mañanas hermos(a) vs. tres mañanas hermosas

A common error is illustrated above. Leaving the Letter "s" off the end does not provide for concordance of number.

* ...tres mañanas hermos [os] vs. tres mañanas hermosas

This example shows a less common error which does not provide for either concordance of number or gender.

[2] Such expansions of the meaning of error is merely a summary and exemplification of the essential grammatical information more fully described in the classic grammars of the specific language.

The capabilities of LINGER-like systems include developing parallel implementations in different languages, even of mixing and inter-mingling different grammars and vocabularies, to create a kind of exploratory learning environment for foreign languages. The new outlook could be effected by these capabilities deriving from LINGER's language independence. Contrastive analysis emphasises the ways language differ from one another. But LINGER-like systems can be developed multilingually and function in a language-independent manner within the context of the Romance Languages. Thus the architecture of LINGER-like systems might one day enable applied linguists to examine the similarities among languages and thus explore new ways of thinking about error. Given the flexibility and polylingual commitments of LINGER, future LINGER-like systems may be the first kind of ITS that are naturally congenial to such a view of language and language learning.

If the instructor has a facility with which he can model the language learning process, it should enhance his ability to diagnose student errors and to refine his suggestions of how the student could avoid them. LINGER-like systems could become an AI-based workbench for the diagnosis of error as a language-independent phenomenon - or as a tool to help in accounting for the differences in particular errors made while learning a target language given a specific native language as the student's starting point. For the teachers, working with language-learning modelling systems would provide an experience which would improve their diagnostic capability. This specific area of application, teacher skill-enhancement through student cognitive modelling, could be a significant new research area for future language instruction.

In conclusion, three answers are available to the question of how one might use future LINGER-like systems: as a linguistic calculator; as an environment for student discovery of new grammatical knowledge; and as a kind of an experimental workbench for teachers to explore the nature of language and the nature of language learning.

Acknowledgments

The LINGER system was developed by Masoud Yazdani and others at Exeter University in England. We are grateful to these colleagues for making available their system, including prototype dictionaries and grammars, and for providing help and guidance throughout this project. Alan Garfinkel has made significant contributions to this paper and the LINGER project at Purdue. The project was supported through an SBIR grant from the U.S. Army to Learning Environments, Inc.

References

Feurzeig, W. 1987
    Algebra Slaves and Agents in a Logo Based Mathematics Curriculum. In *Artificial Intelligence and Education* (Lawler and Yazdani, eds.) Norwood, NJ: Ablex
James, C. 1990
    Learner Language. *Language Teaching* 23, iv, (October): 205 - 13

Keniston, H. 1920
  Common Words in Spanish.
  Hispania 3, 85-96
Moe, A. C. Hopkins, C.J. & Rush, R. T. 1982
  The Vocabulary of First Grade Children
  Springfield, IL. Charles Thomas
Selinker, L. 1980
  Interlanguage
  IRAL, 10:209-231.
Yazdani, M. 1990
  An Artificial Intelligence Approach to Second Language Teaching.
  Journal of AI in Education Vol. 1 No. 3

CHAPTER 7

eL: using AI in CALL

Paul O'Brien

In developing a system that has to possess both an expertise (i.e. knowledge of language structure) and an ability to communicate that expertise (i.e. knowledge of teaching), we have to consider the initial problem from two perspectives - as a computational problem, Can a computer do what we want it to do? and from a pedagogical perspective, how can we use this computational ability and communicate it in a language learning environment ? Naturally, we have already made certain assumptions about the usefulness of our ideal system. The initial hypothesis is based on a pedagogical need. Building an Automatic Syntactic Error Corrector, and finding out how such a system that performs this task can be realised in a learning environment is our initial goal. Involved in the realisation of this goal are two separate criteria one computational, one pedagogical. This chapter will look at the whole system, as well as a few issues related to the design and implementation of a syntactic error correction system.

Design issues

Error analysis: the problem

"The success of error analysis depends upon having adequate interpretations " (Corder, 1981).

It is by assessing the standard of a language learners ability that we can understand the distance between, the existing competence of the learner and the "ideal" one that has to be learnt or acquired. This "ideal" competence is stated either explicitly in a syllabus, or implicitly in the mind of the tutor. A good analysis of learners errors , will produce a good learner model, from which accurate remedial advice can be given. The correct interpretation of the learners intentions is important, if the correct form of remedial advice is to be given.

In CALL, the accurate analysis and correction of learners' language errors could be a powerful tool. This can not only direct tuition, but could be used in the construction of a user model. It is through the accurate assessment of learners that we can distinguish between software that engages the interest of the learner through its becoming more challenging, and that which bores the learner due to its inability to be sensitive to their changing ability.

The analysis of a sentence can be made at different levels with different criterion. For example:

"The hyena ate the house and walk across the Atlantic with its Gills"

The above sentence can be seen at different levels of analysis. We could initially divide our analysis between form and content, and assess its grammatical correctness and its meaning. We could be more specific in our grammatical analysis, by saying it is formally acceptable in word order terms, but unacceptable with regards to agreement (i.e. walk). Referentially it is inappropriate, it has no real world truth and is most probably socially inappropriate (except for a particularly unusual context !). This briefly illustrates that whether a statement is correct or not depends upon what criterion your judgment is based. There can be many levels of analysis, syntactic, semantic, contextual. The knowledge required to make a complete assessment of the sentence , involving all the levels of analysis would be great; firstly, you require a syntactic understanding of sentence structure, including conjunctions, prepositions, etc., and word agreement, secondly, you need some knowledge of semantics so as to "understand" what is being communicated within the sentence, and thirdly, you need contextual knowledge to give this string "real" worldly meaning a context. Such a collection of knowledge, although an aim, is not at present attainable in any reliable form.

Accuracy of error diagnosis, is necessary if any real tuition is to be achieved, a faulty diagnosis can lead to a faulty correction and explanation. This would mislead the learner, and make an error correction system redundant within any educational context. To narrow the scope of system error, and constrain and concentrate the aims of the system, in eL we have addressed primarily syntactic errors, and have not directly looked at semantics.

The general process adopted for error correction is of, firstly, identifying (or diagnosing) the errors in an input sentence, and secondly, correcting those errors by reconstructing the sentence so as to conform to the systems/ or teachers idea of what the sentence should look like.

AI and CALL

Computational linguistics and knowledge representation

" It is ..[the].. representation of knowledge that is the crux of AI" (Hofstadter, 1979)

The specific area of AI this project is mostly concerned with is Computational linguistics ; the processing of natural language by computer. Computational Linguistics can have a powerful role to play in the development of robust, extendible and highly adaptive CALL systems (see Cook and Fass, 1986) . But in order for this role to be fully realised it is necessary to understand the limitations of present day AI and computational linguistic technology, and to steer development of such systems to goals that are attainable within these limitations. It would be unfortunate for CALL to inherit with new AI technology, the "hype" and "exaggeration" that accompanied AI's early development, (Dreyfus, 1963).

Computers have been used to perform many different types of Natural Language Processing (NLP), they can be used as a means of querying a Data-Base (e.g. Chat 80, THEATRE, LADDER) where a language query about a certain topic is answered via an NLP system; they can be used to generate text for conversation (e.g. Eliza, Doctor, Chatterbox), or they can be used for grammatical analysis, (e.g. VP2, OILTS), (Cook and Fass, 1986). Each task requires different types of knowledge and different types of Natural Language Processor as each takes a different perspective on language. The effectiveness of a system depends on how complete, well structured and applicable both its procedural and declarative knowledge is. The way knowledge is represented for a particular domain is the central question.

" .... consider a beach, at the finest resolution, its shape consists of trillions of tiny bumps ... At a coarser resolution it consists of features like dunes ... At the coarsest resolution, it is a ribbon running alongside the ocean. The *right* resolution depends entirely on the use we want to put the object to "(Charniak and McDermott, 1986).

Charniak and McDermott are referring to a perennial question that lies at the root of AI, being what level of representation are we to use for a particular problem. Different representations make different information explicit, pushing other information into the background - the representation chosen has to be the one most suited to the problem domain. In a similar way learners' errors can be seen to occur at different resolutions or levels, we can judge a learner's input at a pragmatic or contextual level, judging its referential suitability or its register; or we can assess it formally, whether or not the input is syntactically correct regardless of its meaning. Therefore, the way of representing the knowledge needed to analyse learners' errors has to be focused at a specific level of analysis. An undertaking to handle all these errors would be foolhardy, as the short history of AI shows, successful systems are built with constrained and focused research goals.

In eL we have aimed at a specific area of analysis, we are dealing with primarily Syntactic Errors, hence our method of knowledge representation has to be adequate primarily for syntactic definition , (if (as it does) this form of representation can have other possible values, then this is a secondary concern). Our representations extend from the definition of morphological structure through to word-order syntax, within this scale of representation we can define certain semantic structures.

Domain and teaching knowledge

One concern that is not normally considered in the development of AI technology, is that our overall goal is not just the modelling of a linguistic process, but the explanation of that process for the purpose of teaching.

We have the knowledge the system uses for analysis (domain) and that which it uses for explanation. In some intelligent systems (e.g. SOPHIE), there is an explicit use of domain knowledge, where the exact reasoning strategy adopted by the system is used to explain its actions to the learner, alternatively, where the domain knowledge is not suitable for explanation, the domain knowledge is implicit, and not seen by the learner, instead another representation of the domain knowledge is used for explanation.

eL has to contain, firstly, the domain knowledge that enables the computer to analyse languages, and secondly, the teaching knowledge of how to explain this knowledge - we are dealing with two different perspectives on language. We have the difference between "grammar" in the sense of defining a formal and complete structure used for computation (as in computational linguistics), and "grammar" as in a teaching device developed by teachers to convey a teaching point. The second is not necessarily formally complete, but a "rule of thumb", used and changed for different learner levels. A language teacher has a more utilitarian approach to language grammar, the basic aim is to communicate an idea; to use it as a device. The computational linguist aims to formally define and codify language .

Therefore, in eL we have to include in our system a certain amount of pedagogical "know-how", through which the systems corrections can be explained. The learner needs to make the necessary inferences between their mistakes and the correction the computer gives. How transparent and lucid the reasoning of the computer is, must be a vital component of a successful CALL system using AI techniques. The representation needed in knowledge -based systems for teaching has to bring to the fore a communicable form of knowledge. In eL we have done this, firstly, by the system generating explanations when certain errors occur; secondly, by a learner interface built in HyperCard (Byron, 1990) which is mainly self-exploratory learning. Ideally, the content of the explanation facilities will be geared towards a specific syllabus, and can present a method of explanation to which the learner is accustomed.

Freedom versus control of interaction

Accuracy of analysis and correction is important in any educational context.

In developing CALL software there is a path to be steered between *freedom* (fluency) and *control* (accuracy) of interaction. The more constrained the interaction, the more accurate the system. Conventional CALL would take a position towards high control of an interaction, where input from the learner is restricted, for example: Choice Master (multiple choice) or Clozemaster . Here, the interaction between computer and learner is highly determined, consisting of a narrow number of input choices the learner can use at each interaction. An AI approach would be moving towards fluency of interaction, where input is comparatively unrestricted, and the learner has more freedom of input. By using AI techniques in eL we are aiming at relatively unrestricted input, the limitations on what input is accepted is the generative power of the knowledge bases. The move towards fluency, not only makes the software more flexible, but extends its "shelf-life" and use. It is more adaptive to the learners' input.

A system that accepts a wide variety of input can have the advantage of being used by different educational activities where language input is required, and also can have a long duration of use. This later solution, carries with it problems of accuracy - so one of the aims is reaching a compromise where within the limits of AI technology we can produce a system that is Fluent (in the sense above) and accurate enough so as to be of educational value. We have to try and maximise accuracy of the system, as well as maintain its comparative flexibility. An understanding of the learner's intentions therefore can be useful in narrowing the scope of misinterpretation. Two major factors that can effect accuracy of an error correction system are firstly, how close the learner's language is to its target form, and secondly, how much the system knows of the learner's intentions, (i.e. the context of use).

- The transparency of the learner's input or how close the original input is to the target sentence is central. The further away the input sentence is to its correct form, the more ambiguous it will be, and the more scope for misinterpretation by the system. Therefore, the accuracy of the system, as with human language comprehension, depends upon how incorrect the sentence is.

- Knowing the overall intention of the learner can narrow the scope if interpretation, narrowing chances of ambiguity and increasing accuracy. By constraining the context of application it is possible to narrow the scope of interpretation,which can be achieved by aiming knowledge bases at specific context goals. This can be achieved by the type of language activity required of the learner, different language activities require different vocabulary, and types of language form. A restaurant would use a specific type of vocabulary, and the interaction with a waiter would be predominantly questions and commands. By adapting the systems knowledge bases for different contexts (for each language activity) it is possible to narrow the scope of misinterpretation. This line of research is presently being developed at the CALL centre at Exeter.

System overview

eL takes learners input and checks it for grammatical correctness. eL was based on a previous system Linger (Language Independent Grammatical Error Reporter) developed by Barchan, (1987). The basic structure adopted in both systems, is that of a classic expert system, with an Inference Engine feeding off Knowledge Bases, and interfacing through a learner interface called *Grammar Points*, (Byron, 1990). In the diagram below , the structure of eL and the learner interface which is shown. Knowledge bases are the stores of linguistic knowledge the "inference engine" uses to check and correct English syntax. The spelling rules and word grammar are permanent as they define the morphology of English and cannot be authored. The *Grammar, Dictionary* and *Checks* are extendible and can be authored. Supplement Grammars, Dictionaries and Checks can be used for specific contexts where for example a particular type of vocabulary would be used.

Fig 1.
eL, and knowledge
bases.

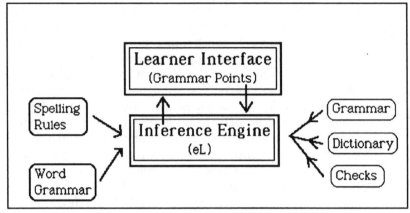

In eL the Learner Interface is developed in HyperText by Byron (1990). The envisaged interaction with the learner involves a hypertext stack, where explanation and language activities are available, this hypertext system is called *Grammar Points* (Byron, 1990). In this interface, it is possible to enter eL from any point, allowing the learner to test the grammaticality of a sentence during an exercise or activity. On leaving eL the learner is returned to the hypertext card he/she had originally been in. eL is accessed through this interface. It is envisaged in future research that language activities in the interface are used to constrain the type of language used, so as to narrow down the scope of ambiguity and misinterpretation during analysis. This will use the Hypertext cards, like "rooms" in a Game Playing scenario, where each "room" represents a context constraining the semantic scope of the language used, although in the case of eL we are mainly dealing with syntax. The overall system would ideally, follow a specific syllabus, at present we have aimed the system at between False beginners to Intermediate. At any point within the eL system (as in *Grammar Points*) a help facility is available to the learner, see Fig 6.

On entering eL you are offered a choice of languages and their supplements. After loading the appropriate language, the learner is asked for a sentence to be typed in (Fig. 2). This is analysed for possible syntactic errors and after analysis output is offered to the learner via a reply option on the menu bar. The learner has access to, firstly, a Result Window (Fig. 3) where eL's version of the learners sentence is presented. Secondly, an Advice Window (Fig. 4) which explains any alterations made to the learners original sentence, and finally, a parse tree (Fig. 5) (or phrase marker) representing the structure of the learner's sentence. Ideally, this information would not be presented in its raw form, but via the explanation facilities which are being developed in HyperCard. This can use the information gleaned from the analysis to explain certain areas of syntax in a much more friendly medium than LPA Prolog, (although when compared to other Prologs, LPA has a commendable arsenal of interfacing techniques).

An example run of eL

In Figs 2, 3, 4 and 5 we have screen dumps of the result of an interaction with eL, illustrating the scope of analysis in eL. There are three areas of correction illustrated, one of strong syntax (i.e. error of word order) and two of weak syntax (sub-word order e.g. agreement). The strong syntax error is where the word "big" is misplaced, and should have been inserted prior to the noun ("man"). The weak syntax errors are firstly, incorrect inflection of the present participle "swim", here "swim" had to be re-inflected with "ing", and secondly, the number agreement between the singular determiner "a", and the plural noun "boats", here "a" is change to "the" . This is the type of information the system has available after analysis, although if needed a certain amount of low level semantics can be gleaned from such a sentence. Fig. 4 shows the advice window used for explaining the analysis to the learner, Fig. 5 the phrase marker or parse tree of the analysis. How much of this information would be made available in the eventual system is a future concern, as has already been said this information will be conveyed via some HyperCard medium, with better explanation facilities.

It is possible for the learner to get a hard copy (print out) of the systems replies and the HyperText cards, for future reference. Also, there is an on line help from anywhere in the system.

Fig 2
Typing in a sentence.

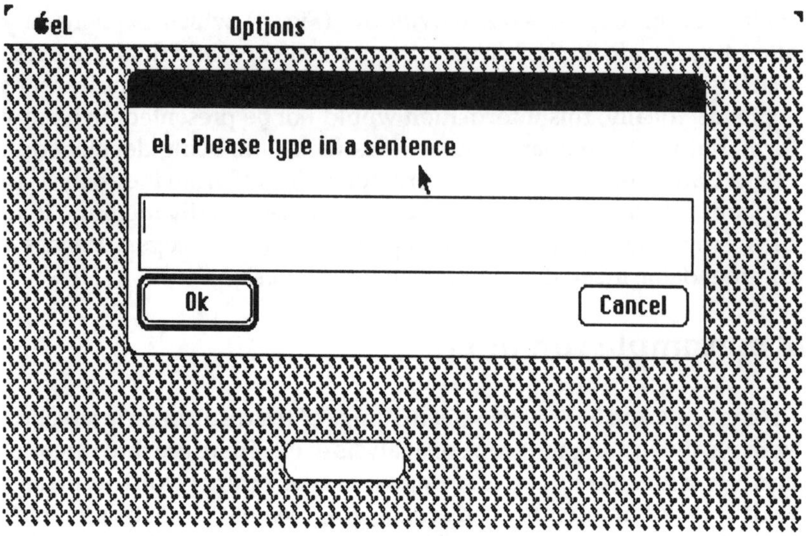

eL   Options

eL : Please type in a sentence

| Ok | | Cancel |

Fig 3
The Result Window.

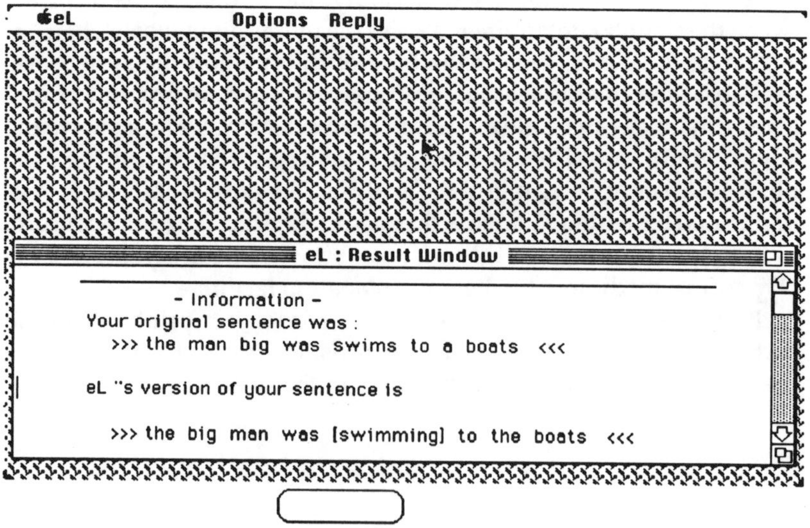

eL   Options Reply

eL : Result Window

- Information -
Your original sentence was :
    >>> the man big was swims to a boats <<<

eL "s version of your sentence is

    >>> the big man was [swimming] to the boats <<<

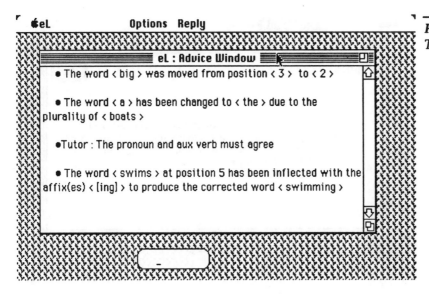

Fig 4
The Advice Window.

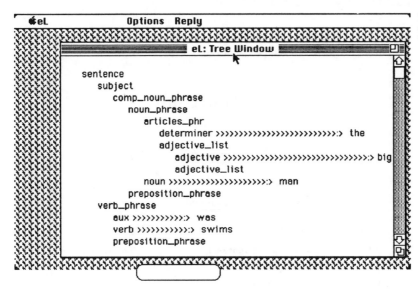

Fig 5
The Tree Window.

Fig 6
The on-line help
facility.

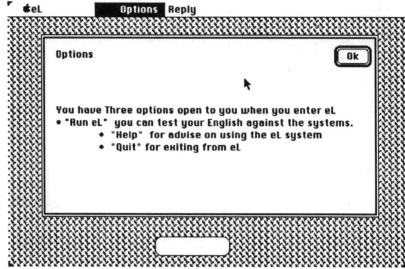

Fig 6
The on-line help
facility.

eL: the system structure

In eL there are three stages to the analysis of a learner's sentence, Preparse, Parse and Post parse stages.

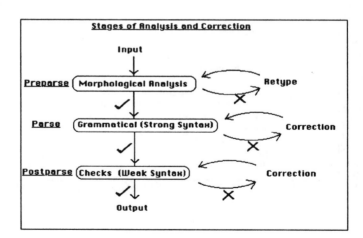

Fig 7
Stages of analysis
and correction.

When the learner types in a sentence, each word is analysed by the preparse stage, if a word is misunderstood , or if the system thinks it is mistyped it will require the learner to re-enter the word. If it is not mistyped, as long as the word is regular, eL will learn the new word by storing it in a personal dictionary . The sentence then enters the parse stage for word order analysis, if this is found to be incorrect the sentence is restructured so as to conform to what eL considers legal. Finally, the input undergoes a post parse stage where agreement, and the correctness of word endings and word suitability is checked. The learner will be then presented with (in the case of errors) a corrected sentence, followed by explanations for each action taken on the   sentence. Each stage will now be looked at in more detail.

The preparse stage

The purpose of the preparse stage is to identify individual words, finding their categories and features. We need to identify a words category (defining its position in a sentence), and its features (defining its structure, for example: first person singular) in order to check the grammatical correctness of a learner's input. This information will later be used in the parse and postparse stage, where the learner's input will be tested for syntactic correctness. In the parse stage only the words categories are looked at, later in the postparse stage their features are analysed. If a word is ambiguous, and can have a number of interpretations - all possible interpretations are carried through to the later stages, and are filtered out in the parse and postparse stage until a final interpretation is achieved. Where words are unidentified in the preparse stage, they are returned to the user for retyping or identification, (See Appendix A run 1).

There are four possible analyses of an input word: direct look-up, where there is direct reference to the dictionary, morphological analysis where the root is known in the dictionary, morphological analysis where the root is unknown and finally, a partial match where the input word is partially matched with an entry in the dictionary. These are all explored in more depth in the following sections.

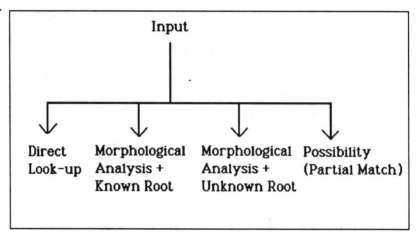

Fig 8
Word analysis.

Direct look-up

Irregular word forms are analysed with direct dictionary look-up. Irregular forms cannot be analysed with a rule-based system (the morphological analyser) as they do not conform to a regular pattern, they are therefore entered individually into the dictionary. When a learner inputs an irregular form this is identified by direct reference to the dictionary.

Root forms are also stored in the dictionary. There are referenced either as uninflected input words, or after an input word has undergone morphological analysis.

Direct look-up involves matching input words with entries in the dictionary, finding a complete match and returning their associated features. The features will contain the syntactic and possibly semantic values the word has, this will be used when analysing the input's syntax, in the parse and post parse stage.

Morphological analysis

" morphology is inherently messy " (Hooper, 1979).

There are two word types we are concerned with in morphological analysis, the lexeme and the grammatical word. The lexeme represents the basic word stem, a pattern that can occur in many forms. For example : "playing, played, plays, player" all share the lexeme "play". They are all versions of the same word, connected semantically in that they share a fundamental meaning, and structurally in that they share a letter combination "p l a y". The lexeme is the basic word form, whereas the grammatical word is the different forms that lexeme can be realised as ("playing", "player" etc.).

A morphological analyser segments a grammatical word into its component morphemes,(i.e. root(s) and inflections) and then checks the legality and effect of this combination. For example :

tries ----> try + s
swimmer ----> swim+er
capability ---> capable + ity

As shown in the example above the system deconstructs the words into their individual morphemes. There two distinct areas of analysis . One area of analysis deals with spelling changes, for example "try" when inflected with an "s" becomes "tries" where the "y" turns into "i", this occurs according to different contexts. Another area of analysis deals with the order of the morphemes and how they combine, this is called morphosyntax or word grammar. For example , the morpheme combination "eat+s+er+ing" ("eatsering"*) is not a permissible combination of morphemes, whereas "capable+ity+s" (or "capabilities") is. Also, combining different morphemes results in different grammatical words, the addition of an "s" to a verb produces a third person singular form (e.g. "walks"), adding an "er" onto a verb can produce a noun (e.g. "swimmer"). We therefore have two stages of analysis, the identification of constituent morphemes and how their spelling changes ; and how these morphemes combine to produce grammatical words (or, inflected forms).

The advantage of a morphological analyser in a Natural Language Processing system is twofold . Firstly, computationally it is efficient, regular forms need only be entered into a dictionary once under its root form. For example : "eat" need only be entered under its root form, inflections and derivations such as "eats","eating","eater" can all be analysed and the root identified, saving the need for multiple entries in a dictionary. Secondly, a morphological component can generate as well as analyse, so if a word is wrongly inflected - the morphological component can inflect it correctly. This aspect has obvious importance when the morphological component is being used with in a syntax error correction system, intended for use by language learners.

The analyser

The morphological component in eL serves the dual purpose of both analysing and generating inflected words. It is based on Koskenniemi's two level model for morphological analysis and synthesis (Koskenniemi, 1983; Russell *et al.*, 1986; Karttunen *et al.*, 1983; Gazdar and Mellish, 1989; O'Brien, 1986). Although the theoretical foundations remain mostly intact, the morphological component in eL differs distinctly in a number of areas from Koskenniemi's system, (these issues are explored later).

The Koskenniemi system uses two-level, bidirectional rules to relate a surface level representation to a lexical level representation. The two-level spelling rules recognise morpheme boundaries and spelling alterations. These rules are bidirectional, permitting both analyses and generation. A word grammar analyses it morphosyntactically defining permissible combinations of morphemes. Basically, different types of spelling and morpheme combinations are defined on two levels. These correspond to a level that represents a segmented word in its individual morphemes (Lexical Level), and a level that represents its inflected (Surface) form. Essentially, the rules "translate" from a string of characters on one level to a corresponding form on another.

Example(i) below illustrates correspondences between Lexical and Surface levels, the rule shown is for the spelling change from "y" to "i", as in "tries", "try+s". The rule states that a "y" becomes an "i" when it is preceded by the sequence "t" followed by an "r", and when it is followed by an "s". Similarly there is a rule for "flies" - "fly+s". These rules are not realistic, as they would over generate, and do not stipulate an accurate enough context, the actual rules would be far more complex. But it gives an idea of how the morphological systems spelling rules operate.

In Fig. 9 the example, representing the rules we have a simplified graphical representation of how the spelling rules are implemented as finite state transducers. The numbers represent states, 1 being the start state, 5 a final state. Each arc has a condition attached to it. For example, to go from state 1 to 2, the input string has to be either a "t" or "f". If the sequence starts with a start state and ends with a final state then the parse is legal. (y.i), signifies the two levels, lexical and surface, stating a correspondence between the two letters.

This example, over generates by permitting the word "tlies" or "tly+s", this is not a flaw, but illustrates the necessity to have a word grammar. The word grammar would take the morpheme combination "tly+s", and with a root and morpheme dictionary, check firstly, if the root and morpheme exist, and secondly test the legality of the morpheme combination in the word grammar.

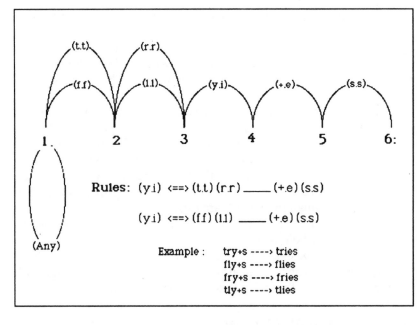

Fig 9
Spelling rules for eL.

A possible word grammar rule could be the following:

```
cat(verb) + fix(suf),cat(s)  ==>
cat(verb),person(3),no(sing)
[cat = category, no = number, sing = singular, suf = suffix ]
```

This word grammar rule states that a verb when combined with a suffix `fix(suf)` and an `"s"` morpheme, produces a third person singluar verb. (Note: the actual formalism form stating word grammar and spelling rules is stated more accurately in the section entitled *Authoring the knowledge base*.)

Although the rule structures and the use of finite state transducers have followed the Koskenniemi model, the morphological component has been altered to suite the particular needs of the present system. The morphological component had to conform to the possibility of misspelt roots, this meant the morphological component could not rely totally on the dictionary to guide its spelling rules. Therefore during word analysis, if a words root cannot be identified by either the dictionary or via a partial match, the system infers what the root could be, from the morphosyntax of the word. eL also parts from Koskenniemi's approach by analysing a words morphology from right to left, instead of left to right. This allows the system to analyse a word ending before its root, thus narrowing the search space by restricting possible root combinations.

The morphological component enables eL to hold knowledge of both what morphemes can combine and how their spelling is affected by such combinations. If a student inflects a word wrongly, eL could illustrate the correction to the original input by showing the word grammar rules that reconstructed the word and the type of spelling re-organisation that occurred. Such knowledge will prove useful when the systems explanation facilities are expanded.

As you would have noticed, we are dealing with inflectional and derivational morphology, not with compound morphology although the system could be extended to cope with this (Koskenniemi, 1983). The present system does not deal with prefixes or infixes, although with alteration to the spelling rules this could be achieved.

Morphological generation

The morphological component also serves as a generator for inflected forms when a word has been identified as badly inflected at the weak syntax stage. This is possible due to the bidirectional nature of the spelling and word grammar rules. When a word is identified as wrongly inflected at the weak syntax stage, the morphological component is served with the root and its necessary goal attributes, the fully constructed word is generated via the word grammar and the spelling rules. Basically the procedure is the reverse of analysis, although the dictionary is involved to a lesser extent.

Example:

" All the *cat* are dead "

The noun "cat" is singular and should be plural. The postparse stage will state a requirement that "cat" (noun-singular) should be made into a pluralised noun. If there is not an actual entry in the dictionary of root "cat" with these required features, then the Morphological generator is used to inflect "cat".

Firstly, it finds an affix (or affixes) that combined with the root and will produce the ideal features, secondly, using the spelling rules it will inflect the noun with this affix (affixes) .

Example:

```
cat(noun),no(sing) +  affix(s)    -->
cat(noun), no(plural)
cat                 +s            -->cats
```

Unidentified words

If eL fails to interpret a word the user inputs but is able to morphologically analyse the word, it will guess the word's root, based on information it holds of the inflections. For example : an "er" inflection can be an inflection of a verb, which could be a candidate analyses for the unknown root.

Possibility match

If a word is input by the learner, and the system fails to identify it either in the Morphological component, or by direct look-up, eL attempts to partially match the word with its dictionary. This is done by a small algorithm which will find partial matches with dictionary entries, working on the percentage of match (match_limit) the user's word has with similar words in the dictionary. This percentage can be changed, but presently stands at 50%. When the learner inputs a word that is not recognised, it allows eL to return a request for the user to retype it suggesting some possibilities. (See *Appendix A, run 1.*)

The predicate command for changing the percentage matched is set_match.

Technical information

The preparse stage results in entries into the Prolog Data Base (Dynamic data entries) recording the analysis of the words the learner/user has input. They are :

lex_root/4: Which states the root form of each input word. This is used as the basis for re-inflection later in the postparse stage.

lex_string/4: States the general analysis of the input words, recording how they were analysed (i.e. either by direct look up (1), morphological analysis and known root (2), morphological analysis and unknown root (3), and possible match(4)).

possibility/4: States all words that have been partially matched with a word in the dictionary, it also holds the percentage matched.

it_is/3: Is the data structure which holds most of the other information, and is used to form the basis for the chart.

The main predicates used are:

pre_parse/1: calls a read-in routine (Clocksin and Mellish), and will analyse input, via either direct look up, morphological analysis, possible match.
parse_word/4: performs only morphological analysis.

go_generate/3: performs only morphological generation.

set_match/0: sets percentage of word to be matched in partial matches.

Reading

For further or background reading see Koskenniemi (1983); Russell *et al.* (1986); Karttunen *et al.* (1983); O'Brien (1986).

The parse stage

The parse stage receives the input sentence as a string of categories from the preparse stage, and checks whether they occur in the correct order. If the sentence structure is correct (i.e. it is permitted by the grammar), a parse tree will be the output. If the sentence is structurally incorrect, a correction algorithm is used to remedy this so as to attain a correct structure.

The parser uses one knowledge base, the grammar. This is a context free grammar, and is used to define permissible word orders. To control what is considered a correct parse, the user would have to author the grammar knowledge base (see *Authoring the knowledge base*).

Chart parsing and the choice of parsing method

The basic parsing method is a top-down, left to right chart parser. Briefly, a chart parser holds a record of syntactic structures which can be combined to produce larger structures - it maintains a record of structures looked for or discovered during a parse. This method is widely used in Natural Language Processing circles, and is suited to our task as it constructs temporary structures which can be used in strong syntactic correction. Knowledge of where a parse fails can aim remedial action at the area of a sentence where the parse failed, and where the structural fault originates.

In eL the output from the preparse stage goes to form the initial chart, (see below). The initial chart is used as the basis for the parse, as the syntactic structures that form a sentence are based on these terminal constituents. Where a words category is ambiguous, each possible interpretation of the word, is given the same vertices. This is disambiguated when a successful parse has been achieved where the category used in the final structure is taken. In the example below "man" is seen to be ambiguous, in that it could be either a noun ("the man ..."), or a verb ("They man the boats"), resulting in a shared chart edge (2 to 3), which will be disambiguated later when the sentence is parsed. This is achieved because the combination determiner + verb, is unacceptable, whereas that of determiner + noun is acceptable - and so the later is selected. Where there is syntactic ambiguity and more than one parse is permissible, both parse structures are output, (e.g. " the man saw the girl with a telescope").

The initial premise for a top down parse is that it is looking for a sentence, hence the active edge from vertex 1 to 1, "sentence?". The parsing process follows that of a conventional chart (tabular) parsing method . Providing the word order is acceptable to the grammar, the chart parser will output a tree structure of the sentence. If the sentence is badly structured or incomplete, a partial tree structure will be produced, this will be used to direct sentence re-organisation, (strong syntax correction). The parser only "sees" categories at terminal vertices, it has no knowledge of the actual words, these are later analysed in the Postparse stage (see *Postparse stage*).

Fig 10
The eL Chart Parser.

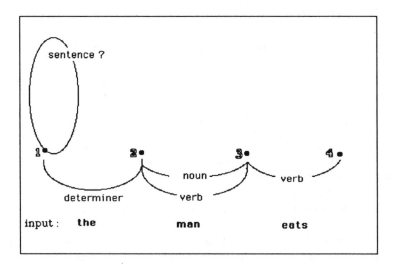

The grammar

The grammar used is a context free grammar with a non-terminal parent node consisting of a combination of terminal and non-terminal nodes. The notation and structure of the grammar is described in more detail in the section *Authoring the knowledge base*.

Why a chart parser ?

The choice of using a chart parsing algorithm and a chart as our data-structure, evolved after using various other methods. The main con-tender was a Definite Clause Grammar (DCG) based parsing system, which was implemented but proved unsatisfactory.

Initially, Linger used a DCG method , which would output a parse tree if the parse had been successful. If the parse failed there was no output from the parser - resulting in a "blind" strong syntax Correction algorithm. (see Barchan, 1987). The first step in remedying this was building a meta-interpreter for the DCG Parsing mechanism (see Pereira and Shieber, pp 159). The meta-interpreter itself was written as a DCG, maximising its efficiency. By controlling the parse at this level, partial tree structures, identified by the depth of parse (level in parse tree), were produced. These, if the parse had failed, would be used to re-organise the word order of the learner's sentence. The basic correc-tion algorithm used information from the parse to identify the position of the parse failure. A set of candidate parents above the failure point were taken, it was to these structures that the input words had to conform. Based on the failure position a string of categories around the failure point were selected. The system would generate possible strings of categories from the parent nodes and match each generation with the actual categories in the input string. A record of how much movement, deletion and insertion was kept giving a score for each re-organisation. The score with the least alteration was selected.

For Example :

Fig 11
Example of sentence failure.

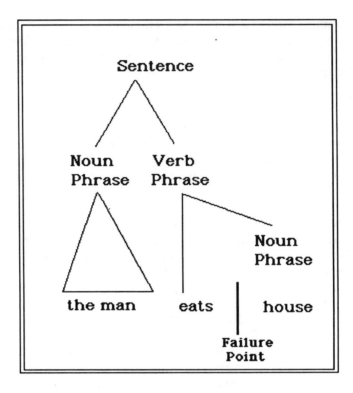

In Fig 11 the failure occurs after the verb "eats". From this failure point, the system would selected candidate parent nodes (i.e. *Verb Phrase*: the closest parent node that is a non-terminal) and generate possible category strings to be matched with the input sentence. This would produce a list of possible remedies, depending on the grammar, one of which will be chosen to correct the input string.

The drawback with this approach was the search space this approach produced, and how it expanded proportionately with the grammar.

- The meta interpreter necessitated a great deal of tree searching, through the form of its output. In order to identify partial trees the depth (level) of the structures had to be recorded - enabling the system to distinguish different structures with the same names.
- The search space for strong syntax correction increased proportionally to the size of the grammar, and could prove prohibitive with a large grammar.

Basically, this method attempted to get a DCG to perform like a chart parser. It was therefore decided to implement a chart parsing based approach. The advantage of this approach is that :

- The parsing process will produce partial and complete tree structures, holding a record of both the parse and what hypotheses were made during the parse.

For example: it will hold information of what categories were searched for and where.

- The input sentence and tree structures were represented via vertices (i.e. positions denoting position of word/structure in the sentence) not by depth of tree which proved cumbersome in the meta-interpreter.
- The parsing algorithm is efficient ; only searching for structures once, where higher nodes in a tree can share sibling nodes.
- The parser can hold two or more syntactic interpretations of a syntactically ambiguous sentence.

Strong Syntax Correction

The final version of eL uses the chart parser, and also adopts a different approach to strong syntactic correction, due to the search space problem of the previous method. In eL if the word order of the user's input is incorrect, it will attempt to restructure the sentence, based on how the parse failed. The overall aim, both in strong syntax and weak syntax correction is the maintenance (if possible) of the users original input. eL is only concerned with misplaced words, major phrasal re-organisation is beyond its capabilities as is phrasal idiomatic correction.

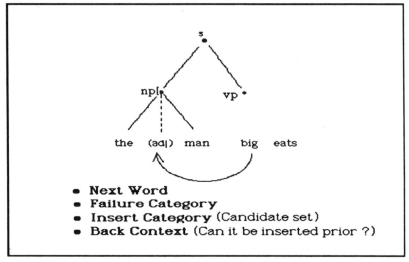

*Fig 12
Strong syntax
correction.*

- **Next Word**
- **Failure Category**
- **Insert Category** (Candidate set)
- **Back Context** (Can it be inserted prior ?)

If a parse fails, the tree structure constructed up to the failure is used to decide the type of correction needed.

A section of the input string is taken to be considered when strong syntax correction is performed. The context used for correct strong syntax correction is the *failure category* (the actual category where the parse failed), the *previous word to the failure category*, the *following word* from the failure, and the possible *insertion categories* which could replace the failure category, these are used to decide what action is to be taken on an input sentence. It is possible to take a context which includes the following words and categories to the faiulure point, because due to the preparse stage the system has knowledge of what categories are coming up in the parse. Due to this context most actions are limited in their scope to the immediate context of the failed word, as a result only local re-organisation is possible, large chunks of structure cannot be manipulated. But words can be moved over the whole length of the input sentence.

The actions possible are the insertion of a new category, deletion of the failure word, movement of a word, or insertion of a word previously deleted. If a word is deleted, it becomes stored as a deleted item and eL attempts to insert it later in the parse so as to maintain as much of the original sentence as possible, the semantic side effects are not considered.

The correction algorithm

The correction algorithm is a list of possible operations performed on the failed parse. One will be performed at a time, and will be fired if their conditins are met. They are coded in this order: 1. being the first tried, 2. the next and so on.

1. Insert a deleted item and try to carry on parse if the deleted item which can be inserted at the failure point.

2. Delete failed word if the following category would enable the continuation of the parse (i.e. the following category is equal to one of the insertion categories) and try and re-insert prior to failure point.

3. Insert the insertion categories if there is no following word, and no failure category (i.e. the sentence is missing something on the end). Although all insertion categories are inserted, whichever one fits the final parse will be the one eventually selected. These are therefore filtered out in the rest of the parse.

4. Delete failure category and try and insert it in the sentence prior to the failure point.

5. Delete failure category.

6. Insert the insertion categories.

For Example:

*Fig 13
Context for strong
syntax correction.*

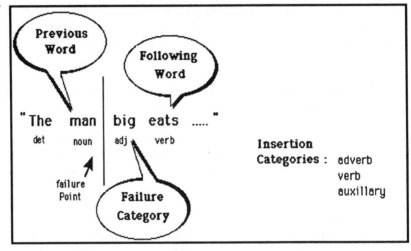

The insertion categories are those categories that could legally replace the failure category. In Fig. 13 the categories "adverb","verb" or" auxiliary could be placed in the "big" position.

In Fig. 13 where the adjective "big" is situated in a wrong position, "big" is the failed word (adjective its failure category), the following word "eats", is also of the same category as the insertion categories. Therefore we know that by deleting "big" we can proceed with the parse, also , because of the retention of active edges in the chart we also know it is possible to re-insert an adjective prior to the noun, therefore based on this , "big" is moved to word position 2. (This is procedure 2 of the correction algorithm.)

Improvements and drawbacks

- To increase parsing efficiency and help in strong syntax correction a bidirectional chart parsing approach might prove useful. (De Roerk and Steel, 1988).

- To maintain as much of the original sentence as possible, this is not necessarily the best method. But considering the lack of any high-level semantic information is the best method available, with the level of analysis.
- The strong syntax correction algorithm cannot deal with phrasal idiomatic language - this is considered a must by some language teachers.

Technical information

Main dynamic data structures used:

chart/6: records the parsing strcutures, inlcuding vertices, parent node being searched for, unfound parts of parent node being searched for, Tree holding the strcuture found so far.

Main predicates used:

call_parse/0: (parse/0) calls the main parsing algorithm.

strong_syntax_correction/0: calls the main correction algorithm for strong syntax (or word order correction).

The postparse stage

The postparse stage is our second stage of analysis/correction. At this point in the analysis we are concerned with weak syntax, having checked the word order of the learner's sentence, the system goes on to analyse the actual word structures, their features and their syntactic suitability. The words are analysed for their suitability and agreement, teaching points can be picked up here. The postparse stage feeds off one main knowledge base that is the *Checks* knowledge base (see section on *Authoring the knowledge base*).

The basic analysis mechanism is quite simple, and the knowledge bases are flexible and can be authored to perform quite powerful operations. The checks are essentially, context-dependent production rules. Stating a *context* (i.e. syntactic structure) in which they are applicable, a list of *conditions* (what needs to be true for the rule to fire), and a list of *actions*, which are subsequently applied to the sentence (see below).

in <context>
if <conditions>
then <actions>

The conditions and actions are lists of condition/action primitives which can be used to test the learners input and act on it. A list of these are in the section on *Authoring the knowledge base*.

There are two stages in the analysis:

1. Check applicability of the rule

2. Call the actions if the rule is applicable

When a rule is selected from the knowledge base. Initially its context is checked. A rules context is a syntactic structure, defined in the grammar. Each rule states the structure in which its actions are applicable, if this structure is found in the parse tree then the rest of the rule will be performed on that structure. If the context structure exists in the parse tree, the structure is found, and the list of conditions executed on that structure. Once the conditions have been satisfied as to the rules applicability, the action part of the rule is called.

The action part of a rule is taken in two parts - *State requirements* and *Act on those requirements*. The action list is fired sequentially. Some actions have no actual effect on the input sentence, but are used for explanation, for example: comments can be coupled with checks, these comments will be output to the learner at the end of the analysis.

The actions state a list of requirements which need to be true in the stated context, with the stated conditions. If these requirements are not true in the input sentence, an ideal feature list is constructed stated the required features. These ideal feature lists are then acted upon changing the learners input so as to conform to the requirements of the rule.

Each rule is acted on in sequence, hence the importance of ordering the checks. It would be wrong to change one word for another, after an agreement rule - as the later rule should be executed on the final word used.

Below are a list of example checks to illustrate the expressive capacity of the checks and their primitives. The checks below are given in a "pseudo prology" manner to give you more feeling for the way the checks are designed and used. Although the syntax of the example checks is close to the actual form, inorder to make them easier to read the odd bracket etc. have been omitted, for a formal definition of structure see *Authoring the knowledge base.*

Agreement

Example:
"he walk" ---> "he walks"

```
in      sentence
if      exists(verb)
        exists(pronoun,noun_phrase)
then    agree(verb,pronoun,[person,plurality]),
        comment(['The pronoun and verb
        must agree']).
```

The above check, states that *in* a sentence (Context), *if* there exists a verb in a verb phrase and a pronoun in a noun phrase (conditions), *then* the pronoun and the verb must agree in the features person and plurality (actions). This check will apply to all similar structures in the learners input. It must be noted the close relationship, firstly, between the checks context and conditions and the Grammar (via the parse tree), and secondly, its relationship to the Dictionary through it referencing data regarding the words feature values. A degree of conformity has to be maintained when constructing these knowledge bases. It is necessary to know what features you require at the postparse stage when constructing the grammar and dictionary.

Teaching points

Example:
" he enjoy to swim" --> "he likes to swim"

output = *In the infinitive phrase it might be better to replace "enjoy" with "like"*

```
in      verb_phrase
if      has_features
        (verb,[ root(enjoy) ])
        exists_str(infinitive_phrase)
then    change_word(enjoy,like),
        comment(['In the infinitive
        phrase its might be better to
        replace "enjoy" with "like"'])
```

This check is very specific changing the word "enjoy" to "like" when using an infinitive phrase. It states that *in* all `verb_phrases` *if* there exists a verb with the root "enjoy" and there exists an infinitive phrase, *then* change the word ":enjoy" to "like" and make the comment "In the infinitive etc.". This check would over generate in certain contexts, but illustrates the access the check has to feature lists of input words via such primitives as `has_features()`. The `comment()` primitive will output this statement when the result of the analysis is given to the learner.

Semantic marking

Example:
"Hi there John "
output = *The greeting "Hi" was given in a(n) informal manner to John*

```
in      greeting
if      has_features(greeting,
        [manner(Variable),
        root(Variable2)]),
        has_features(name,
        [root(Variable3)])
then    comment(['The greeting"',
        Variable2,'"was given in a(n)',
        Variable,'manner to',Variable3])
```

The above check illustrates how it is possible to pick up certain semantic markings a word is awarded in the dictionary and use it later. The checks states that *in* a greeting (which we presume is a syntactic structure), *if* there exists a greeting which has the features manner and root, and there exists a name with a root : the values of these are the instanciations of variables 1/2/3, *then* give the comment "The greeting etc." with the appropriate variables inserted.

These three examples illustrate the type of syntax and power of the checks mechanism. The checks author has a variety of primitives to use, with which conditions and actions can be stated. A possible pedagogical way of using the checks is by recording with each entry in the dictionary a feature value (translation) which holds the translation of the English word into the native tongue of the learner. This could be picked up at the postparse stage and output to the learner, in whichever way the checks author sees fit.

The agreement action makes use of the morphological generator, where words can be inflected with affixes and their spelling altered to suit. (see Morphological Generation: in the section on *The preparse stage*).

Semantic marking II

Example:
"Le homme est ..."
output = *"homme" means "man" in English*

```
in      noun_phrase
if      exists(noun),
        has_features(noun,
        [root(X),trans(Y)])
then    comment([''"',X,'"means"',Y,'"
        in English']).

word([h,o,m,m,e],[cat(noun),
..... trans(man)],[]).
```

The above example illustrates how any type of feature value can be added in the dictionary, to be picked up in the postparse stage. Here a feature trans() is added to the dictionary entry of a French word ("homme"), this states a word for word translation. This is picked up in the above check, and returned to the learner.

General comments

Generally, the check formalism is flexible enough to allow the knowledge base author to design the checks to suit their own specific purpose. As you can see it is still a very "Prology" language, unfortunately an authoring system has not been built. When building the checks it shoiuld be noted that it is a combination of dictionary entries and the feature lists, the grammar to define contexts and structures, and of course the checks to define the actions to be taken.

The checks are fired linearly, and so their ordering is important (see *Authoring the knowledge base*). The checks can be rationalised by making the conditions as specific as possible. Thus preventing over generation.

Technical information

Main Dynamic Data Base entries:

ideal/2: stating the ideal feature list for a certain word in the input string.

comment/1: stores comments to be made when the answer is output to the user.

Main Predicates that are called:

check/0: calls the post parse stage

post_parse/0: calls action to be taken on the word once the requirements have been stated in ideal().

Authoring the knowledge base

[1]In the context of eL Language will refer primarily to English, unless otherwise stated.

eL has two types of Knowledge Base (KB), *extendible* and *permanent* (see Fig. 14). The knowledge bases can be seen as files which contain facts about language[1] . They contain the systems "knowledge", hence the name "knowledge bases". The permanent knowledge bases are not authorable, as they address English morphological spelling and morphosyntax, areas of syntax which are completely defined, and need not be altered nor extended. The extendible KB's are dynamic as they can be authored to address various areas of English Syntax. The extendible knowledge Bases can be authored outside the system via MacWrite.

eL feeds off three extendible "knowledge Bases", They are:

- The Dictionary
- The Grammar
- The Checks

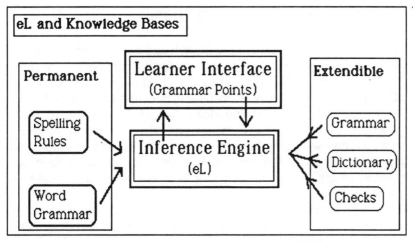

Fig 14
eL and knowledge bases.

Each contains information about a certain area of language structure, and each is used when analysing a learners sentence. They can be altered (hence the name "dynamic") for different areas of language, allowing the possibility of having various sets of KB's (i.e. grammar,dictionary,checks), each addressing different areas of language. For example : A lesson can be set around the size of animals in a zoo ; the dictionary could contain the language that would be used in such a subject area, and the grammar and checks could be altered for the use of comparative and superlatives, creating a set of knowledge bases that could be used to teach comparatives and superlatives within the context of a zoo.

" The elephant is bigger than the chimpanzee "

This approach uses the knowledge bases to target language at specific contexts, narrowing down scope for ambiguity, misinterpretation, and incorrect - correction; and widening the possibilities of semantic marking. It allows the system to constrain the context of application. This approach necessarily goes hand in hand with the design of the language activity the learner is required to perform. The activity has to be steered towards the contexts defined in the knowledge bases.

Authoring the knowledge bases

When authoring the extendible knowledge bases of eL it must be remembered that the three knowledge bases are interdependent and so an alteration in one knowledge base can effect the working of another, all the knowledge bases therefore have to be consistent with each other. Prior to any editing of the knowledge bases it is advisable to have a general idea of what areas of syntax you wish to define. With this in mind it is advisable that the author has a basic procedural understanding of the way eL analyses and corrects grammar, this will colour their method of authoring and improve both the speed and quality of authoring.

The preferred process of authoring , although this need not be followed, is to construct primarily the grammar and dictionary; once the word order has been defined and tested, the checks can be written. This sequence is because the checks , use the structures and features expressed in the dictionary and grammar.

The syntax of the knowledge bases

The eL knowledge bases are Prolog code files, and so should conform to certain basic syntactic requirements of Prolog. A brief summary of the conventions are listed, although these will become clearer later in the examples.

Prolog Conventions:

- Each complete entry in the knowledge bases must be terminated by a full stop.
- Each entry must be in lower case, except when using variables in the checks.
- A List structure is ended by an opening and terminating square bracket, and each entry within it is separated by a comma.

Grammar

The grammar is written as a context free grammar, inheriting all the expressive limitations this suggests. Essentially, the grammar should express word order ; the permissible combinations of words your knowledge bases wish to accept. A close relationship exists between the dictionary and the grammar, this is due to the parsing process. To explain briefly, the preparse stage uses the dictionary to define the words in terms of categories, this is then used by the parser to identify structure - therefore, all the parser sees of the input sentence are the words possible categories.

The grammar is expressed as rules. These have a right-hand side (RHS) and left-hand side (LHS), and are divided by an arrow ('==>').

```
<Parent Category> ==> <Constituent Elements>
```

example:
```
sentence ==>[subject, predicate].
subject ==>[name].
predicate ==>[verb].
```

The above example states that a sentence consists of a combination of a subject followed by a predicate. And a subject is a name, and a predicate a verb. Such a grammar, providing the dictionary awards the category *'name'* to "John" and *'verb'* to "eats", would accept "John eats" as a legal sentence, as defined by this grammar.

Dictionary

The *Dictionary* is used by both the checks and grammar, and so is of primary importance when building the knowledge bases Each word entry consists of the word entry, followed by a list of its features and the words referent (see later). The list of features must contain the words category, and can also contain whatever features are necessary or required in the checks. Any amount of new features can be introduced into the dictionary entries, which can be picked up and used by the checks, it is here that the system could use a limited set of semantic markers.

Example:

syntax: `word(<WORD>,<Feature List>,<Referent>)`

example: `word([e,a,t],[cat(verb),`
          `subcat(reg),tense(pres) ],[eat]).`

syntax: `general_cat(cat(<subcategory>,`
          `<list of possible feature lists>).`

example: `general_cat(cat(reg),[`
          `[person(1),plurality(s)],`
          `[person(2),plurality(s)],`
          `[person(1),plurality(p)],`
          `[person(2),plurality(p)],`
          `[person(3),plurality(p)]]).`

In the above example, you have the word entry for "eat", which has the features of a "category" called "verb", in the present tense. The `"subcat(reg)"` allows the entry to inherit features common to all regular verbs, hence it inherits the features from `general_cat()` "third person singular","second person singular" etc. The features can be invented by the author, it is possible to use a disjunctive "or" between feature values, for example "cat(verb or auxilliary)". The referent entry , is merely an identifier, identifying a group of related verbs that share the same root. This is used mainly with irregular forms.

For Example:

```
word([a,m],[cat(verb or aux) etc........],
[be]).

word([i,s],[cat(verb or aux) etc........],
[be]).

word([a,r,e],[cat(verb or aux) etc........],
[be]).

word([b,e],[cat(verb or aux) etc........],
[be]).
```

In the above example, "be" is a way of relating the irregular forms of the verb "to be". When the system has to correct subject verb agreement this enables the system to identify related lexemes.

Checks

The checks have three sections, the context, the condition, and the actions. The context defines the grammatical structure in which the check is to be applied, this is always a left-hand side of a grammar rule. The conditions define what needs to be true in order for the action part of the rule to be executed, and the actions effect some action on the parsed sentence if the context and conditions have been met.

```
in <Context>
if <Conditions>
then <Actions>.
```

Each part has a set of primitives which can be used to define either the context, conditions or actions. The primitives are the means by which any condition or action can be expressed, obviously the primitives for the actions and those for the conditions are different as they are required to perform different tasks. You will find that they can be used to express quite complex structures, and can even extend to certain semantic marking.

It is necessary to consider the order of the checks in the knowledge base. There is a distinct hierarchy involved in how the checks are ordered. For example : word changes should precede any alteration to the word itself. Therefore when authoring the checks the user should be conscious of the hierarchy:

```
insert word
delete word
change word
agree
(not)have_features
comment
```

The check primitives

Context: Any structure defined in the grammar.

Conditions:

`exists(category)` [e.g.`exists(noun)`]:there exists a category "noun" in the context stated.

`not_exist(category)` [eg.`not_exist(noun)`]:there does not exist a category "noun" in the context stated.

`exists_str(structure)` [e.g.`exists_str(noun_phrase)`]: there exists a structure "noun_phrase" in the context stated.

`not_exists_str(structure)` [eg.`not_exists_str(noun_phrase)`]: there does not exist a structure "noun_phrase" in the context stated.

`exists_str(structure,category)` [eg.`exists_str(verb_phrase, verb)`]: there exists a structure "verb_phrase" with a category "verb" in the context stated.

`exists_str(structure,category,word)` [eg.`exists_str(verb_phrase, verb,eat)`]:there exists a structure (verb_phrase) consisting of a category "verb" that is the word "eat" in the context stated.

has_features(category, features)[e.g.has_features(verb, [root(like),person(3)])]:the category "verb" in the context has the features "root(like)" and "person(3)".

not_has_features(category, features) [e.g.not_has_features(verb, [root(like),person(3)])]: the category "verb" in the context does not have the features "root(like)" and "person(3)".

Actions:

agree(Category1,Category2,feature) [e.g. agree(pronoun,verb,[person,plurality])]:within the context the "verb" has to agree with the "pronoun" in " person" and "plurality".

have_features(Category, features)[e.g. have_features (noun,[root(man)])]: withinthe context the noun has to have the features "root(man)".

not_have_features(category, features) [e.g.not_have_features(noun,[root(man)])]:within the context the noun must not have the features "root(man)".

change_word(Word1,Word2)[e.g.change_word(man,boy)]: within the context the word "man" must be changed to the word "boy".

change_word(category, position,features) [e.g.change_word(noun,2,[person(3),plurality(p)])]: within the context the "noun" in position "2" must be changed to any word that has the features "person(3)" and "plurality(p)".

change_word2(category,word) [e.g.change_word2(noun,boy)]: within the context the category "noun" must be changed to the word "boy".

Semantic marking

Certain semantic markers can be traced through the dictionary, through the grammar in to the checks, so as to give some indication of the meaning of a sentence. The checks pick up semantic-feature values which can be tested for in the conditions and used for informed output, see the section on *The postparse stage.*

Spelling rules and word /morpheme grammar

Morphological analysis and generation uses two permanent knowledge bases for analysis and generation - the spelling rules and word grammar. The spelling rules state how a word's spelling will be affected when it is inflected with certain affixes, the word grammar will state what combinations of morphemes are permitted with in a language. (For more information see *Preparse stage.*)

(Note: these knowledge bases are not expected to be authored, their syntax and structure is complex. Prior to any authoring read some relevent literature on the Koskenniemi system.)

Spelling rules

The spelling rules state correspondences between a surface and lexical level. Essentially, morphological analysis/generation is moving from one level to the other. Each rule state contexts in which certain spelling alterations occur when certain morphemes combine. The spelling rules as seen in the section *preparse stage,* are represented in eL as finite state transducers. Ideally, a compiler would be constructed to read the Koskenniemi formalism and produce the transducer knowledge base.

```
move(State 1,State 2, Surface Form --->
Lexical Form).
```

Surface Form := SA|no([SA])|or([SA, ...
SAn])|subset/set of SA

Examples:

> `move(1,2,a--->V)`: A member of the alphabet
> Surface Alphabet (see below)
> `move(1,2,no([x])--->V)`: The surface character can be
> any but "x"
> `move(1,2,or([a,b,c])--->V)`: The surface character can
> be either "a", "b" or "c"
> `move(1,2,vowel--->V)`: The surface character can be any
> member of the set "vowel"
> `move(1,2,0--->+)`: Do not consume any input string, no
> conditions attached to surface string.

Lexical Form := `LA|[morpheme name]|Prolog Variable`

Examples:

> `move(1,2,a --->V)`: A member of the alphabet Surface
> Alphabet "a" becomes an "a" on the lexical level
> `move(1,2,a --->b)`: The surface character"a" becomes a "b"
> on the lexical level
> `move(1,2,x --->[ent])`: The surface character is consumed
> and the morpheme of name "ent" is starting

Surface Alphabet (SA) :=
`{a,b,c,d,e,f,g,h,i,j,k,l,m,n,o,p,q,r,s,t,u,v,`
`w,x,y,z,0 }`

Lexical Alphabet (LA) :=
`{a,b,c,d,e,f,g,h,i,j,k,l,m,n,o,p,q,r,s,t,u,v,`
`w,x,y,z,0,+,[label] }`

The spelling rules define sequences of letters at both a lexical and surface level, and how their spelling changes when combined with different morphemes. The move() predicate represents a state transition in the finite state transducer, stating one correspondence of a lexical and surface string. Shown above is the move() predicate and its associated syntax.

The `move()` predicates represent sequences of states and correspondences that can be made is a complete transition is made from a start state to a final state. A sequence is defined by stipulating a succession of states.

For Example:

```
move(1,2,a --->x).
move(2,3,b --->y).
move(3,4,0 --->+).
move(4,-1,c --->z).
```

The above example states a sequence from state 1 , via states 2,3,4 to -1. With a string of surface characters "abc" (with a null string at states 3 to 4) , corresponding with a lexical level "xy+z".

Examples:

Using Null String and Morpheme Boundary

```
lying     lie+ing

move(1,2,l--->V).
move(2,3,y--->i).
move(3,4,0--->e).
move(4,5,0--->+).
move(5,6,i--->V).
move(6,7,n--->V).
move(7,-1,g--->V).
```

The correspondence between "lying" and "lie+ing" on a lexical level means that we cannot have a one-to-one correspondence between the letters of one level to the other. The null string permits the insertion of a character in its corresponding level, without there being any consumption of the string in that level. In the above example, an "e" and "+" are inserted on the lexical level without any consumption of the surface form "lying". In the above example a morpheme boundary is inserted on the lexical level.

Using "or" and sets

Example:

```
swimmer      swim+er
bigger       big+er

move(13,or([b,g])--->V).
move(1,1,any--->V).
move(3,4,i--->V).
move(4,5,or([m,g])--->V).
move(5,6,or([m,g])---> +).
move(6,7,e--->V).
move(7,-1,r--->V).

any([a,b,c,d,e,f,g,h,i,j,k,l,m,n,o,p,q,r,s,t,u,v,
w,x,y,z ]).
```

In the above example we have a case of gemination or consonant doubling. The state transition from state 1 to 1 (i.e. move(1,1,any---->V)) will consume any member of the set "any" defined above. This will consume the front of words until a letter "b" or "g" occurs, where the first move would be appropriate. This move() uses the definition or() which allows a disjunction to be stated. The order of the move() clauses is important as they are chosen linearly and the most specific first (e.g. a correspondence (a.b) is more specific than (any.any)).

Word grammar

```
List of features and List of affix features =>
Resulting List of features
```

Example:

```
[cat(verb)|_] and [cat(er)|_] =>[cat(noun)|_].
morph(er,[cat(er),root?(verb)]).
```

The word grammar states what combination of morphemes is legal and how their features change when combined. In the example above we have the combination of a "verb" morpheme and an "er" morpheme which when combined becomes an grammatical word of category "noun". The morphemes are held in a morpheme library in the predicate `morph()`. This consists of the name of the morpheme ("er" above) and a list of its features. `root?()` states a possible category for the root, based on the morpheme. In the example, an "er" ending would most likely be affixed to a "verb", this becomes useful if the identity of the root is not known.

Conclusions

Problems

Some problems with the eL system which should be addressed are:

1. It cannot cope with phrasal idiomatic language, which is vital if the system is intended to address language used within situation contexts (e.g. restaurant).

2. Its speed of processing, is dependent on the amount of errors in the input, with some strong syntax errors processing time can be a lot, resulting in the breaking of the learners attention. Making a faster implementation of the present system would be invaluable in a classroom context. Similarly, the speed of the system is proportional to the size of the Knowledge Bases. Again necessitating a faster implementation. Areas such as Bidirectional chart parsing (De Roerk and Steel, 1988) should be explored for a faster parsing algorithm.

Future development

When the system is authored the knowledge bases should be targeted at specific contexts, this narrows scope of system error, misinterpretation, ambiguity - making the system more viable in an educational context.

The knowledge bases can be authored so as to aim at specific contexts, aiming the dictionary, grammar, etc., at specific teaching points with different themes. This follows the idea used in adventure game idea design where each "room" has its own context, language, etc. This would especially work well in the Hypertext media.

eL is a system still under development at Exeter University, that combines traditional computer science with AI techniques. The domain of application has been constrained so as to increase accuracy. It combines domain knowledge with, pedagogical knowledge. At present the system must still be seen as a prototype, after further integration of the inference engine (eL) with the learner interface (Byron, 1990), and the development of specific knowledge bases aimed at specific contexts - then the system will be ready for classroom testing.

Features to note are the strong syntax correction algorithm, the flexibility and power of Checks and the morphological component, each with possible applications independent of the overall system.[2]

Acknowledgments

I would like to acknowledge the support of my colleagues Masoud Yazdani and Don Byron. This research is sponsored by the Open Learning Branch of the Training Agency, and directed by Masoud Yazdani.

References

Barchan J. 1987
Linger: A Language Independent Grammatical Error Reporter
Mphil Thesis Department of Computer Science , Exter University
Byron G. 1989
eL; A Tool for Language Learning, *Computer Assisted Language Learning* Vol. 2 pp 83-91.
Charniak E. and McDermot D. 1986
Introduction To Artificial Intelligence
Addison-Wesley, Reading, Massachusetts
Cook V.J. and Fass D. 1986
Natural Language Processing by Computer and Language Teaching System Vol. 14 No.2 pp.163-170
Corder S.P. 1981
Error Analysis and Interlaguage
Oxford University Press

[2]*System Specifications* eL has been implemented on an IBM compatible (386 machine) in Prolog2 as well as on a Mac SE/II using LPA Mac Prolog with a HyperText Interface.

131

De Roeck, A. and Steel, S. 1988
Bidirectional Chart Parsing
Cognitive Science Centre, Department of Computer Science,
University of Essex.

Dreyfus R. 1963
What Computers Can't do : A Critique of Artificial Reason
Harper & Row Publishers

Gazdar G. and Mellish 1989
Natural Language Processing
Addison-Wesley

Hofstadter D.R. 1979
Godel , Escher, Bach : an Eternal Golden Braid
The Harvester Press Ltd.

Karttunen *et al*. 1983
Kimmo : A General Morphological Processor.
Texas Linguistic Forum No. 22.

Koskenniemi 1983
A Two Level Model for Morphological Analysis IJCAI'83.

O'Brien P. 1986
A Review and Implementation of Koskenniemi's Morphological
Analyser / Generator. MSc Thesis. Computer Science Department,
University of Essex.

Russell, Ritchie, Pulman, and Black 1986
A Dictionary and Morphological Analyser for English,
Coling 1986.

Thompson H 1981
Chart Parsing and Rule Schemata in
*GPSG Proceedings of the 19th Annual Meeting of Association for
Computational Linguistics,*
ACL Stanford, California.

Appendix A: example runs

The next section includes a sequence of example runs to illustrate certain features of the system. Most of the examples are taken from the Macintosh version of eL although some come from the IBM/Prolog2 version - hence the differing user interfaces.

Example run 1: Illustrates the partial match algorithm of the preparse stage and the pre-parse stages limited capacity to learn regular word forms. This example is taken from the IBM version of eL.

Example run 2: Illustrates Strong syntax error recovery and the use of insertion.

Example run 3: Illustrates Strong syntax error recovery and the use of insertion and movement (i.e. deletion and insertion).

Example run 4: Illustrates Strong syntax error recovery and the use of insertion. Here the system inserts at the end of the sentence.

Example run 1: IBM/Prolog2 version of eL

Figs 15 and 16 illustrate an example run of eL on the IBM.

Fig 15
*An illustration of
the partial match
algorithm.*

```
yes
?- go.


        Welcome to eL : please type your sentence after the
prompt

        ==>Th man walk to the garden.

            th * man * walk * to * the * garden *
        _____


        eL has had difficulty understanding < th >
        eL has guessed it might be "the" or "to"

        Is it :
            a : Mistyped
            b : A new word
                > a
        Please re-enter word
                the
                the *
        _____


        eL has had difficulty understanding < garden >

        Is it :
            a : Mistyped
            b : A new word
                > b
        eL will need to learn the new word
        Please enter the words root
            => garden
        Please enter the words Category(s)
        ==>noun

        The new words root is "garden" and its category is
"[noun]"
            Is this okay ? (y or n)
                => y
                garden *

        Input : the man walk to the garden
```

```
Input : the man walk to the garden


    ..... parsing

sentence
    subject
        comp_noun_phrase
            preposition_phrase
            noun_phrase
                articles_phr
                    pre_det
                    determiner ›››››››››››››››››››››››:› the
                    adjective_list
                noun ›››››››››››››››››››:› man
    verb_phrase
        verb ›››››››››››:› walk
        preposition ›››››››››››:› to
        noun_phrase
            articles_phr
                pre_det
                determiner ›››››››››››››››››››:› the
                adjective_list
            noun ›››››››››››››››:› garden


        ********* PRESS RETURN *********



_____

            - Information -

Your original sentence was :
    ››› th man walk to the garden  ‹‹‹
The correct version of you sentence is :

    ››› the man [walks] to the garden  ‹‹‹
==================================================

Tutor : The subject ‹ man › and the main verb ‹ walk
› should agree
The word ‹ walk › at position 3 has been inflected with
the affix(es) ‹ [s] › to produce the corrected word
‹ walks ›
```

Fig 16
The weak syntax agreement correction.

135

Fig.15 illustrates the partial match algorithm in the preparse stage, where the system suggests possible interpretations for the input word that has not been recognised. It also, illustrates how eL can learn new word entries if it does not recognise a word, providing the word is a regular form.

Fig. 16 presents the weak syntax agreement correction and comments made.

Example run 2: strong syntax correction: insertion - Mac eL runs

Fig 17
Example run of strong syntax correction: insertion.

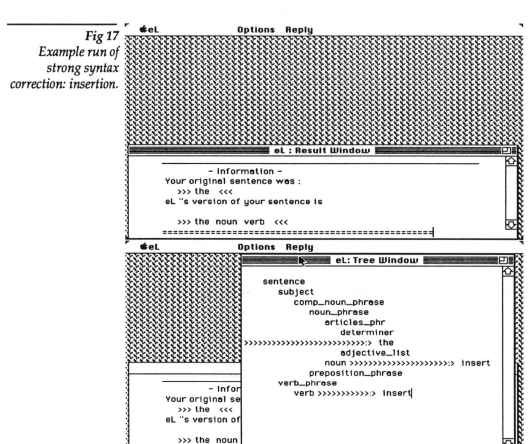

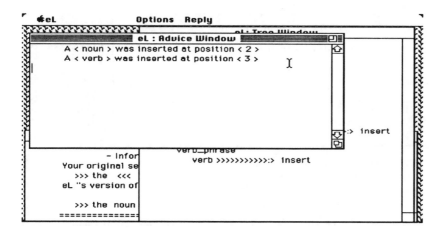

The above example run illustrates eL's recovery from two strong syntax errors. The incomplete sentence "the" has been completed by the insertion of a noun and verb.

Example run 3: strong syntax correction - insertion and movement

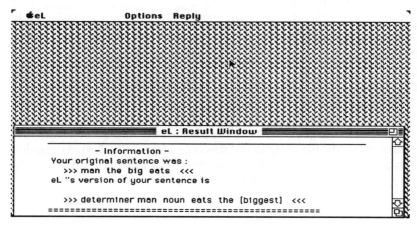

Fig 18
Example run of strong syntax correction: insertion and movement.

*Fig 18 (continued)
Example run of
strong syntax
correction: insertion
and movement.*

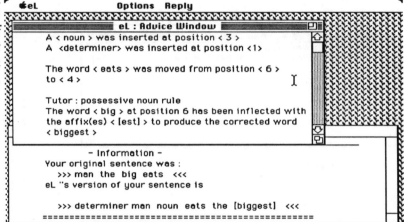

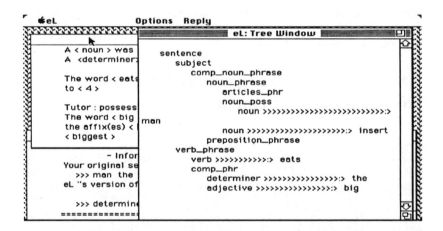

The above example illustrates how a severely incorrect sentence can result in a radically reorganised sentence. How the system re-organises such input depends on the elasticity of the grammar, and how the sentence can be accommodated within it.

Example run 4: strong syntax correction

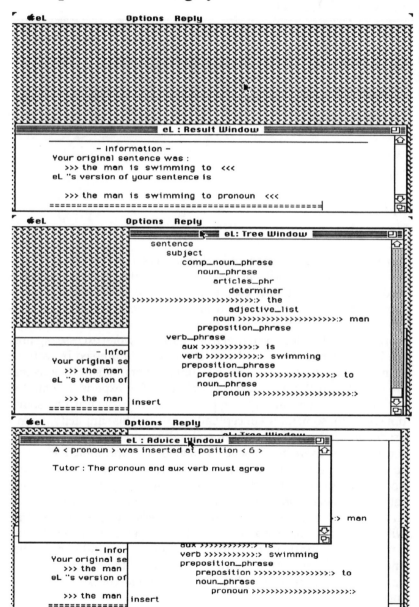

Fig 19
Example run of
strong syntax
correction at the end
of the sentence.

CHAPTER 8

Grammar checking programs for learners of English as a foreign language

Philip Bolt

As a lecturer in a department of English one is sometimes asked if there is some program that will check the grammaticality of sentences that a student writes and that is easy to operate and understand. What is sometimes sought is a program or environment that will act as a grammatical 'cocoon' and that will check and correct word order and word form and protect the writer from his or her worst excesses against the English language. In addition, students unimpressed by various games, manipulation exercises, gap-filling programs, simulations that are intended to stimulate real discussion or after experience with the various applications of 'hyper' type ideas to language learning some-times look for something conceptually simpler. This simplicity, from a user's perspective, and the desire for something more dynamic, context-sensitive and more responsive to their own unprompted input, is sometimes accompanied by a wish for more direct program interven-tion and advice on text produced by learners often in a word processing setting. "Is there a program I can use to help me with my writing of single sentences - not one that will help process my ideas, not one that will tell me the outline of a business letter and not one that will tell me if my writing is too dense or simple according to some index, but rather one that simply checks that the right words are in the right place." "We have spell checkers why can't we have grammar checkers ? " we are sometimes asked.

We have of course had spell checkers for some time and very useful they can be, although not completely reliable given the existence of homonyms and keyboard 'errors' that nevertheless result in words e.g. form and from. Spell checkers deal only with individual words and their operations are inspired to an extent by typical keyboard error in addition to the linguistically motivated legal letter sequence information. Grammar (and style) checkers also exist. Grammar, in the more constricted sense of syntax, however, concerns words in combination and in relation to each other and is a much more complex phenomenon - as attested to by the many research programs devoted to aspects of natural language processing. It is not too unfair to say that the principles of word combination - especially when the meaning(s) and message(s) carried by such combinations are considered - are far from understood to a degree sufficient for comprehensive and consistent embodiment in a computational application, at least on the sorts of machines to which most of us and our students have access. Nevertheless, as evidenced by the seven programs reviewed here, attempts have made to bring some of the fruits of this developmental work to smaller machines and some of these efforts end up as grammar and style checkers, the notion of 'checking' possibly being analogous with the 'checking' of spell checkers.

Program architecture

Most grammar and style checkers consist of a large dictionary of words and some way of setting these words into patterns that represent a structure for each sentence - a parsing process - and a way of presenting either the structure itself, or observations based thereon, to the user with appropriate comments.

The dictionary - finding the word and its categories

The use of the term 'dictionary' is somewhat metaphorical as they generally contain part(s) of speech/word category information about each word and, depending on their level of development, other information about other relevant morphosyntactic features, e.g. plurality, number and countability, rather than providing pronunciation information, meanings and a range of examples. The dictionary provides each word with a part of speech or grammatical category marking and, although words are generally considered in isolation, most of the programs provide for words belonging to more than one word class, e.g. 'man' as noun, verb or adjective.

Parsing - putting the word categories into syntactic structures

Setting the words, or word/part of speech categories into patterns - phrases, clauses and sentences - is the second principal phase for grammar checking programs. Despite attempts to arrive at a workable description of sentence use and meaning based on 'higher' level categories of meaning, purpose and function in addition to an initial or immediate appeal to syntactic categories and form (Jackson, 1990; Brown and Miller, 1991; Allen, 1987), it remains the case that most comprehensive and practical computational applications have in fact to address word class assignment and combination issues at a fairly early stage. The parsing process, or processes, that define and identify appropriate combinations of grammatical categories within individual sentences, and which transform the user's input into the system's representation(s) of sentence structure, are the most crucial elements in grammar checking systems. It is these parsing and syntactic structure identification capabilities of word combination and sentence representation which drive the system, they are the "heart of the system" in the sense that they identify the possible structure(s) of a sentence. Through these processes a structure (or structures) for a sentence is either found or, if one cannot be found, this fact is, or should be, reported to the user.

Presentation - showing the user what has been done

From the combination of word class assignment and the sentence representing process, and the resulting sentence structure (or lack of such), the program is then able to inform the user of any problems encountered and how they may be put right. The accuracy and value of such information depend crucially upon how well individual words have been identified, how well syntactic structures have been constructed and how well information about acceptable word combinations are integrated into the resources of the system. If a grammar checker (as with spell checkers) cannot recover an accurate representation of the structure of their object of attention - individual sentences at the very least - then, like an ill-informed tutor, it would appear that the program is in no position to offer considered, context-sensitive comment.

The review

This review considers seven programs by examining their performance against a small but considered set of sentences which contain errors of a syntactic/morphosyntactic nature. In addition, where possible, we seek to determine how the heart of each system works, that is how (and how well) each program represents the syntactic structure(s) of a sentence and from this representation, how (and how well) they differentiate the well formed from the ill-formed sentences and comment informedly on the latter. We start by describing the seven programs and consider the rationale for the tests made in relation to one class of users and their likely errors. We then look at the results of matching the sentences containing certain error types against the seven programs. Finally, we consider in brief the architecture necessary for an adequate grammar checker and to what extent this is realised in the programs it is possible to examine. In so doing, the value of the use of the analysis made by each system - its dictionary, parsing and presentational resources, especially the first two as they do or do not inform the third - will be considered not only from the perspective of the target end user but also from the perspective of someone wanting to adapt the program. In this way it will be possible to consider the systems both from the perspective of their inherent capabilities and future potential and sub-categorise them in terms of their transparency and potential.

The programs

The choice of the programs examined: *Correct Grammar for Windows* (a particular development of *CorrecText*) [1], *Right Writer for Windows* [2], *Grammatik for Windows* [3], *CorrecText* (as implemented in Complete Writer's Toolkit) [4], *Reader* [5], *PowerEdit* [6] and LINGER [7] - reflects the fact

[1] *Correct Grammar for Windows*
Lifetree Software Inc., 33 New Montgomery St., Suite 1260, San Francisco,USA, Writing Tools Group Inc, Houghton Mifflin Co.

[2] *Right Writer for Windows*
QUE software, 11711 North College Avenue, Suite 140, Carmel, Indiana, USA

[3] *Grammatik for Windows*
Reference Software, California USA

[4] *Reader*
Langsoft Ltd. 5 Rosemont Road, London W3 9LT UK.

[5] *Poweredit*
Artificial Linguistics Texas, USA

[6] *CorrecText Grammar Correction System*
Houghton Mifflin Company, Language Systems Inc.

[7] *Linger*
Department of Computer Science, Exeter University, UK

that they are available to various degrees for use on small computers and are claimed to offer grammar checking on such systems. Doubtless these seven do not exhaust the number of such programs, certainly of such programs under development, but they represent a cross-section of the well-known, the commercially available and those under development. The resources needed to run these five programs are slightly different - *Grammatik for Windows, RightWriter for Windows* and *Correct Grammar for Windows* run on a standard IBM PC, with Microsoft *Windows 3, CorrecText, Reader* and *PowerEdit* run under DOS on a standard PC, and LINGER runs on a standard IBM compatible, but to be effective requires a 386 based machine, *Windows 3* and at least four megabytes of memory.

Correct Grammar, Right Writer, Grammatik, CorrecText, Reader and *PowerEdit* all run effortlessly on the standard IBM PC with most word processors. They are mostly file-based in the sense that they take complete files and process them rather than checking individual sentences. *PowerEdit* will analyse individual sentences and *Correct Grammar* allows the use of the Clipboard for passing individual sentences. They will accept any input but comment only on some. Although in addition to checking for grammar these programs also offer advice on style - jargon, clichés, redundancy, archaic or colloquial use, weak wording and long-windedness - together with 'mechanical' type features - spelling, punctuation and balancing of brackets and parenthesis or quotation marks and in most cases the ubiquitous readability indexes - our focus will be on the grammatical devices as defined and implemented by each program. LINGER checks only individual sentences.

The seven programs claim to offer advice on the following relevant grammatical features:

Correct Grammar: The use of A/an, many/much, pronoun usage, this/that/these/those, confused verbs, confused possessives, confused homonyms, confused phrases, sentence fragment and sentence run-on.

RightWriter: The occurrence of noun-verb mismatches, sentence fragments, run-on sentences, misused verbs, pronoun and article usages, wrong verb forms, possessive use, misused word and a more general - 'is this <item or pattern> correct?'.

Grammatik: The occurrence of incomplete sentences, possessive forms, homonyms, pronouns, number agreement, verb agreement, article and incorrect verb forms.

Reader: The uses of a/an and <GRA??> - a general grammatical error marker.

PowerEdit: The uses of a/an, this/that/these/those, subject-verb agreement, words that modified nouns, forms of pronouns, coordination and a general - words that 'do not belong together'.

CorrecText: The occurrence of 'ungrammatical expressions', noun phrase consistency, mass Vs count/a Vs an, verbal group consistency, subject-verb agreement, pronoun errors and clause errors.

LINGER: The occurrence of a/an, a Vs an Vs some, many/much with countable/uncountable nouns, this/that Vs these/those, pronoun usage in single and embedded clauses, subject-auxiliary-main verb agreement and possessive forms.

LINGER - Language INdependent Grammatical Error Reporter - is a system that has been under development at Exeter University for some five years, Barchan (1987), Bolt (1988), Yazdani (1990) and O'Brien (1989), and as such it has all the features of a system under development. It is written in Prolog and will accept single sentences including those with words that are not currently in its dictionary but which can be incorporated by the user at run time. It will accept a wide range of sentences but is able to parse and correct only those for which it has the relevant structures and resources - these, however, are extendible. In general it can be very accurate for simple sentences but becomes increasingly erratic when dealing with longer sentences.

The users

The seven programs do not all address the same audience but claim a range of the facilities that might well appeal to various types of user. The six commercial programs are intended for the user with little time to spend on writing tasks and for writing in general and who are writing for a business context. As such, a lot of their attention is directed towards getting the user to eschew particular uses of English, promoting, rather, an economy of style of expression which is perceived to be favoured in the business world and which emphasises action and 'positiveness'. Nevertheless, the improvement of writing skills and proof reading are also stated aims of these programs:

> "*Right Writer* is useful if your second language is English. We have added many new rules that will help find common mistakes you make." "*RightWriter's* unique artificial intelligence uses two ultra-modern technologies - parsing and expert systems - to evaluate your documents."[8]

> "*Grammatik* is much more than a grammar checker. It is a sentence analyst, style and usage guide, spelling checker, readability analyst, and proofreader . . . " and "Whether you are a corporate manager, a government official, a novelist, a professor, or a student, *Grammatik* will help you to communicate more effectively."[9]

> "*Reader* is a software tool for writers. It checks the grammar, punctuation, spelling and word usage of any kind of English text, including literary, business and technical genres."[10]

> "*PowerEdit* simulates the reactions of a human reader. *PowerEdit's* messages tell you when a reader will stumble over your syntax or will quickly absorb your message."[11]

> "The *CorrecText* Grammar Correction System is a powerful, easy-to-use program that helps you produce grammatically correct documents."[12]

[8] *Right Writer for Windows Guide* pp2-4, and Right Writer for Windows box cover.

[9] *Grammatik for Windows User's Guide* p1.

[10] Documentation accompanying *Reader*.

[11] *PowerEdit User's Guide* p1.

[12] *The Complete Writer's Toolkit* manual p35.

so that we might reasonably expect them to be able to handle the simpler sentences of the language learner. For LINGER, the intended user is the foreign language learner of English at the beginning and lower intermediate levels of competence and the range of sentences it can handle reflect this.

The principal user identified in this article is the learner of English at a fairly low level of competence and who might understandably view such programs as offering the kind of dynamic and context-sensitive facility that might help with the identification and correction of grammatical error as and when it occurs. The use of the seven programs as grammar checkers at whatever higher levels of competence presupposes the capability of accurate checking at the lower levels - there would seem to be no hot line to the higher level facts of syntactic/ stylistic/textual structure that can reasonably be accessed without addressing, or registering at least, those that are in error at the 'lower' morphosyntactic and syntactic levels that is, if a program cannot identify the problems in a sentence such as "Peter was feels tired." when it is an independent clause, it is unlikely to be able to identify the problem when these four words function as, for example, part of a subordinate clause.

The errors

The types of error used here are not culled from any particular first language context. Rather, they represent errors made by different students in different places and at different times. More importantly, however, they cover the basic resources/error problems that any successful grammar checking program should be able to accommodate. Such an accommodation would go some way to providing the grammatical coccoon noted above, at the lower levels of competence at least.

The tests

The central problems for all the programs are the definition and recovery of the structure of the user's input, input in which the structure might by no means be readily apparent. The program must be able to represent the structure of the input in terms of the clauses and phrases that go to make up the sentence. At the lowest level, therefore, this is analogous to one major objective for vision systems, finding the 'edges' of the objects which go to make up a whole scene (Boden, 1988; Marr, 1982). As with vision 'processing', there is an interdependency between the whole scene - the sentence - and the identification and meaning of the objects which make up the scene - the words and the phrases. Later, through the test sentences we examine how well each program can be said to determine the 'edges' of its objects - what it identifies as words, word categories, phrases and clauses - the relationship(s) between the objects that these edges define and the functional role(s) they perform.

Natural language, even if confined to syntactic patternings, is infinite in its range of forms (even more so when the meanings of these forms and their possible interpretations are considered). As such, it would be unrealistic to expect that these programs would be able to deal adequately with more than a proportion of the forms of English, despite the very large dictionaries they use - Complete Writer's Toolkit uses the American Heritage Electronic Dictionary which "contains 115,000 words and word forms",[13] "a comprehensive 500,000-dictionary",[14] "*PowerEdit* contains a comprehensive lexicon of tens of thousands of English words and phrases."[15] Coverage might appear more of a stumbling block for native users and advanced users whose constructions and uses might reasonably be expected to be more complex and creative - in the same way that certain words used by some authors are not readily recognised by some spell checkers. However, for whatever level of user intended, we might reasonably expect that a grammar checker will be able to accommodate the sorts of language which inevitably appear in beginner and intermediate English as a foreign language texts.

Such language comprises a core of grammatical constructions and devices from which the following have been selected:

(a) indefinite articles - 'a' versus 'an',
(b) determiners - 'this' versus 'these',
(c) determiners and quantifiers with countable and uncountable nouns,
(d) subject/verb and subject/auxiliary agreement with simple subjects, including main verb forms,
(e) subject/verb agreement with coordinated subjects,
(f) subject/verb agreement with subjects having post noun phrase and relative clause modification,
(g) possessive forms,
(h) pronouns in object positions,
(i) pronouns in embedded clauses,
(j) modal auxiliaries
(k) forms of auxiliaries and main verbs,
(l) adjectival and verb complementation.

[13] *The Complete Writer's Toolkit manual.* p35.
[14] Documentation accompanying *Reader.*
[15] *PowerEdit User's Guide* pD-1.

These represent only a small range of the possible and actual types of error that the typical user is prone to make. No account is taken of lexical choice or appropriacy of sentence. In addition, errors made are often not as pure as the individual listing suggests - frequently, they involve more than one of the following; a badly chosen word, a word that is out of place - wrongly located, redundant or missing - and often a fusion of constructions. In brief, while certain types of learners may well make certain kinds of errors at certain stages of their development there is in fact an infinite range of possibilities and the above offers a target language characterisation of a very small part of this infinity. The range of sentences used in part two reflect both the pure, single errors that occur and more complex combinations. Finally, to avoid the ever present strictures of the 'readability' and 'strength' indexes contained in most of the commercial programs, some longer sentences have been included.

The performance of the seven programs

The performance of the programs with the representative sentences is set out below. It should be noted that in some cases more than one response and comment and correction and piece of advice was generated. Recorded here is the relevant response for each error underlined. If the program 'processed' the sentence but produced no correction or advice the original sentence is reproduced. The programs are indicated thus; *Correct Grammar for Windows* (CG), *CorrecText* (CT) , *Reader* (RD), *PowerEdit* (PE), *Right Writer for Windows* (RW), *Grammatik for Windows* (GK), and LINGER (LI).

(a) indefinite articles - 'a' versus 'an',

(ai) He writes *an letter.*

Program   Response and/or advice

CG   He writes [- Consider 'a' instead of 'an'. -] an letter.
CT   >Consider [a] instead of {an}.
RD   He writes an<< | GRA: Bad a/an | letter.
PE   <Usag> Should "an letter" be changed to "a letter"? If not, then "very" should be changed in the database.
RW   Should this be "a" instead of "an?" Replace with "a."
GK   He writes <USE 'A' FOR WORDS BEGINNING WITH A CONSONANT SOUND.> an letter.

LI > he writes a letter - a noun like < letter > needs an 'a' not an 'an'

(aii) He writes *an letter new*.

CG He writes [- Consider 'a' instead of 'an'. -] an letter new.

CT >Consider [a] instead of {an}.

RD He writes an<< I GRA: Bad a/an I letter new.

PE <Usag> Should "an letter" be changed to "a letter"? If not, then "letter" should be changed in the database.

RW Should this be "a" instead of "an?" Replace with "a."

GK He writes <USE 'A' FOR WORDS BEGINNING WITH A CONSONANT SOUND.> an letter new.

LI > he writes a new letter - the word <new> which is a(n) <adjective> was moved from sentence position <5> to <4> - a noun like < letter > needs an 'a' not an 'an'.

(b) determiners - 'this' versus 'these',

(bi) *This books* is very old.

CG This [- Consider 'book' instead of 'books'. -] books is very old.

CT >Consider [book] instead of {books}.

RD This books is very old.

PE <Gram> The use of "books" and "is" may be unclear. If this is a subject-verb relationship, then "is" must agree in number with "books." If not, "books" or "is" may be part of a special phrase or may be unclear. Can you clarify? <Usag> "This" does not seem to match "books." Do they belong together? Are they part of a special phrase? Has a word such as "that" been deleted? Is there a missing comma?

RW This books is<<*_ Do the noun and the verb agree in number in This books is? Change the appropriate one to make the noun and verb agree. *>><<*_ Is "This books is" correct? This is unusually worded. Please study it to see if it will be clear to your reader. If not, reword it. *>> very old.

GK This books is very old.

LI > these books are very old

the subject noun before the main verb and the main verb must agree - a plural noun such as < books > needs a determiner lik e 'these' or 'those' not 'this'.

(bii) I want to send *this very interesting books* to him.

CG   I want to send [- Consider 'this' instead of 'these'. -] this very interesting
[- Consider 'book' instead of 'books'. -] books to him.

CT   >Consider [this] instead of {these} OR Consider [books] instead of {book}.

RD   I want to send this very interesting books to him.

PE   <Usag> The words "this very interesting" cannot usually be used together.

RW   I want to send this very interesting books to him

GK   I want to send <USUALLY 'THIS' SHOULD BE FOLLOWED BY A SINGULAR NOUN.> this very interesting books to him.

LI   > i want to send these very interesting books to him a plural noun such as < books > needs a determiner like 'these' or 'those' not 'that'.

(biii) The bear eat *that berries old.*

CG   The [- Consider 'bears' instead of 'bear'. -] bear [- Consider 'eats' instead of 'eat'. -] eat [- Consider 'those' instead of 'that'. -] that [- Consider 'berry' instead of 'berries'. -] berries old.

CT   > Consider [bears] instead of {bear}. OR Consider [eats] instead of {eat}. > Consider [those] instead of {that}. OR Consider [berry] instead of {berries}.

RD   The bear eat that berries old.

PE   <Gram> The subject for "eat" may be unclear. If it is "bear," then "eat" must agree in number with it. The structure of this sentence may need to be clarified. <Disc> There may be a word missing in this sentence. Make sure that there are a subject and verb present in "that berries old," and that it is properly coordinated with the rest of the

> sentence.

RW The bear eat<<*_ Do the noun and the verb agree in number in The bear eat? Change the appropriate one to make the noun and verb agree. *>> that berries old.

GK The <THE SINGULAR NOUN 'BEAR' MAY BE USED INCORRECTLY WITH THE PLURAL FORM OF THE VERB 'EAT'.> bear eat that berries old.

LI > the bear eats those old berries - The word < old > was moved from sentence position < 5> to < 4 > - a plural noun such as < berries > needs a determiner like 'those' or 'these' not 'that' - the subject noun before the main verb and the main verb must agree.

(c) uses of quantifiers and determiners with countable and uncountable nouns:

(ci) He has many evidence.

CG He has [-The word 'many' does not agree with 'evidence'. -] many [- The word 'evidence' does not agree with 'many'. -] evidence.

CT >The word {many} does not agree with {evidence}. ORThe word {evidence} does not agree with {many}. > The word {many} does not agree in number with the following noun.

RD He has many evidence.

PE <Gram> "Evidence" cannot usually have modifying words such as "many" in front of it. <Usag> "Many" does not seem to match "evidence." Do they belong together? Are they part of a special phrase? Has a word such as "that" been deleted? Is there a missing comma?

RW He has many evidence

GK He has <USUALLY 'MANY' IS USED WITH A PLURAL NOUN.> many evidence.

LI > he has much evidence - an uncountable noun like <evidence> needs a determiner like 'many'.

(cii) There is much reasons.

CG   There is [- Consider 'many' instead of 'much'. -] much reasons.

CT   > Consider [many] instead of {much}.

RD   There is much reasons.

PE   <Gram> The use of "there" and "is" may be unclear. If this is a subject-verb relationship, then "is" must agree in number with "there." If not, "there" or "is" may be part of a special phrase or may be unclear. Can you clarify? <Prec> "There" may not be the best subject, especially when used with "is" as a verb.

RW   There is much reasons.

GK   <BE SURE A SINGULAR SUBJECT FOLLOWS 'THERE IS'.> There is much reasons.

LI   > there are many reasons - a countable noun like <reasons> needs a determiner like 'many' - the verb and the determiner and a countable noun <reasons> must agree.

(ciii) An dogs big was in those house

CG   An [- Consider 'dog' instead of 'dogs'. -] dogs big was in [- Consider 'that' instead of 'those'. -] those [- Consider 'houses' instead of 'house'. -] house.

CT   > The word {An} does not agree with {dogs}. OR Consider [dog] instead of {dogs}.
> Consider [A] instead of {An}.
> Consider [that] instead of {those}. OR Consider [houses] instead of {house}.

RD   An dogs big was in those house.

PE   <Prec> Weak words like "was", "an", "those" and "big" do not convey much useful information in this context. Should a more descriptive word be used?

RW   An dogs<<*_ Should this be "a" instead of "an?" Replace with "a." *>> big was in those house<<*_ Is "those house" correct? This is unusually worded. Please study it to see if it will be clear to your reader. If not, reword it. *>>.

GK   <USE 'A' FOR WORDS BEGINNING WITH A CONSO-NANT SOUND.> An dogs big <BE SURE YOU ARE

USING 'WAS' WITH A SINGULAR SUBJECT. ('IT WAS.').> was in <USUALLY 'THOSE' SHOULD BE FOLLOWED BY A PLURAL NOUN.> those house.

LI   > some big dogs were in that house. The word < big > which is a(n) < adjective > has been moved from sentence position < 3 > to < 2 > - a singular noun such as < house > needs 'that' or 'this' not 'those' - a plural noun like < dogs > needs an 'some' not an 'an' - the subject noun before the main verb and the verb must agree.

(d) subject/verb and subject/auxiliary agreement with simple subjects, including main verb forms:

(di)The *man expect* the letter.

CG   The [- Consider 'men' instead of 'man'. -] man [-Consider 'expects' instead of 'expect'. -] expect the letter.

CT   > Consider [men] instead of {man}. OR Consider [expects] instead of {expect}.

RD   The man expect the letter.

PE   <Gram> The subject for "expect" may be unclear. If it is "man," then "expect" must agree in number with it. The structure of this sentence may need to be clarified.

RW   The man expect<<*_ Do the noun and the verb agree in number in The man expect? Change the appropriate one to make the noun and verb agree. *>> the letter.

GK   The <THE SINGULAR NOUN 'MAN' MAY BE USED INCORRECTLY WITH THE PLURAL FORM OF THE VERB 'EXPECT'.> man expect the letter.

LI   > the man expects the letter - the subject noun < man > and the verb must agree.

(dii) *They* is happy.

CG   They [- Consider 'are' instead of 'is'. -] is happy.

CT   > Consider [are] instead of {is}.

RD   They is happy.

PE   <Gram> The use of "they" and "is" may be unclear. If this is a subject-verb relationship, then "is" must agree in number with "they." If not, "they" or "is" may be part of a special phrase or may be unclear. Can you clarify?

RW   They is<<*_ Do the noun and the verb agree in number in They is? Change the appropriate one to make the noun and verb agree. *>><<*_ Is this the correct verb? Replace "They is" with "They are." *>> happy.

GK   They < BE SURE YOU ARE USING 'IS' WITH A SINGULAR SUBJECT. ('IT IS.').> is happy.

LI   > they are happy - the subject pronoun 'they' and the verb must agree.

(diii) *An young dog arent eats* the smelly fish.

CG   An young dog [- Misspelled? -] arent eats the smelly fish.

CT   > The word {arent} is not in Main or User Dictionary.
         > Consider [A] instead of {An}.

RD   An<<|GRA: Bad a/an | young dog arent eats the smelly fish.

PE   <Reln> The use of "an young" may not be correct. Make sure that "an" can perform the action named by "young." If not, then you may need to clarify this sentence. <Reln> The use of "arent eats" may not be correct. Make sure that "arent" can perform the action named by "eats." If not, then you may need to clarify this sentence.

RW   An young<<*_ Should this be "a" instead of "an?" Replace with "a." *>> dog arent eats the smelly fish.

GK   An young dog <SINGLE-WORD SPELLING ERROR.> arent eats the <AN ADVERB ('SMELLY') DOES NOT USUALLY MODIFY A NOUN ('FISH'). YOU MAY NEED TO USE THE ADJECTIVE FORM OF 'SMELLY' (E.G., 'QUICK' INSTEAD OF 'QUICKLY').> smelly fish.

LI   > a young dog isn't eating the smelly fish - an adjective like < young > needs an 'a' not an 'an' - the subject noun before the auxiliary verb and the auxiliary verb must agree - if the word 'isn't' 'aren't' or 'weren't' is a negative auxiliary verb then the following main verb must be in the present participle form -- 'writing'.

(div) *The dog has eat* the bone so it *cant being* hungry.

CG	The dog has eat the bone so it [- Consider 'cants' instead of 'cant'. -] cant being hungry.
CT	> Consider [cants] instead of {cant}.
RD	The dog has eat the bone so it cant being hungry.
PE	<Eleg> "So" tends to be overused. Could you use a word that is more specific or descriptive?
RW	The dog has eat the bone so it cant being hungry.
GK	The dog <IF 'HAS' IS BEING USED AS AN AUXILIARY VERB, IT SHOULD BE USED WITH A PAST PARTI-CIPLE VERB (E.G., 'HAS FIXED').> has eat the bone so <IT IS UNUSUAL TO FIND A NOUN OR ADJECTIVE SUCH AS 'CANT' FOLLOWING THE PRONOUN 'IT'. YOU MAY MEAN THE POSSESSIVE FORM OF THE PRONOUN, OR NEED A ',' OR A VERB.> it cant being hungry.
LI	> the dog has eaten the bone so it can't be hungry - if the word 'couldn't' 'can't' 'wouldn't' or 'shouldn't' is a modal auxiliary verb then the following main verb must be in the base form --- 'be' - a modal verb in contracted form must be written 'can't' or 'mustn't' - if the verb 'has' 'have' or 'had' is an auxiliary verb then the following main verb must be in the past participle form -- 'written'.

(dv) The *dog didnt eating another fishes.*

CG	The dog [- Misspelled? -] didnt eating another fishes.
CT	> The word {didnt} is not in Main or User Dictionary.
RD	The dog didnt eating another fishes.
PE	The dog didnt eating another fishes.
RW	The dog didnt eating another fishes.
GK	The dog <SINGLE-WORD SPELLING ERROR.> didnt eating <USUALLY 'ANOTHER' SHOULD BE FOLLOWED BY A SINGULAR NOUN.>· another fishes.

LI   > the dog didn't eat another fish - the subject noun before the auxiliary verb and the auxiliary verb must agree - if the word 'doesn't' 'don't' or 'didn't' is a negative auxiliary verb then the following main verb must be in the base form -- 'write' - a determiner like 'another' or 'each' must have a singular noun like 'book'.

(dvi) Although *they was* very tired they persuaded the *manager to leaves*.

CG   Although they [- This verb group may be in the passive voice. -] was very tired they persuaded the manager to leaves.

CT   > This verb group may be in the passive voice.
     > Consider [were] instead of {was}.
     > This sentence does not seem to contain a main clause.

RD   Although they<< | STY:Passive | was very tired they persuaded the manager to leaves.

PE   <Gram> The use of "they" and "was" may be unclear. If this is a subject-verb relationship, then "was" must agree in number with "they." If not, "they" or "was" may be part of a special phrase or may be unclear. Can you clarify? (Choppy flow) This sentence consists of many small parts. The essential parts may be difficult to find. Can you clarify?

RW   Although they was<<*_ Is this the correct verb? Replace "they was" with "they were." *>> very tired<<*_ Do the noun and the verb agree in number in they was very tired? Change the appropriate one to make the noun and verb agree. *>> they persuaded the manager to leaves.<<*_ Is this a complete sentence? If so, is there a comma missing? RightWriter cannot find the part of the sentence that is not conditional. There is no comma separating the conditional part of the sentence from the rest of the sentence. Either insert a comma where appropriate or complete the sentence. *>>

GK   Although they <BE SURE YOU ARE USING 'WAS' WITH A SINGULAR SUBJECT. ('IT WAS.').> <PASSIVE VOICE: 'WAS TIRED'. CONSIDER REVISING USING

ACTIVE VOICE. SEE HELP FOR MORE INFORMA-
TION.> was very tired they persuaded the manager
to leaves.

LI   > although they were very tired they persuaded the
manager to leave - the subject pronoun 'they' and the
verb must agree - the verb in an infinitive clause must
be in the base form.

(dvii) A *good book for she to reading are* under the desk.

CG   A good book for she to reading [- Consider 'is' instead of
'are'. -] are under the desk [- Consider deleting the
space before this punctuation mark. -] .

CT   > Consider [is] instead of {are}.

RD   A good book for she to reading are under the desk.

PE   <Prec> Weak words like "are, she or good" do not convey
much useful information in this context. Should a more
descriptive word be used?

RW   A good book for she to reading are under the desk .<<*_
Should there be space before this   punctuation? *>>

GK   A good book for <THE CONTEXT OF 'SHE' INDICATES
THAT YOU MAY NEED TO REPLACE IT WITH THE
OBJECT PRONOUN 'HER'.> she to reading < BE SURE
YOU ARE USING 'ARE' WITH A PLURAL SUBJECT
('WE ARE', 'JILL AND DON ARE').   SINGULAR
SUBJECTS JOINED BY 'OR' OR 'NOR' USUALLY TAKE
A SINGULAR VERB ('JILL OR DON IS').> are under
the desk.

LI   > a good book for her to read is under the desk - the
pronoun in an embedded clause needs to be of the form
--- them her or him - the subject noun before the main
verb and the verb must agree - the pronoun in this
clause needs the accusative form --- him her or them.

(e) subject/verb agreement with coordinated subjects,

(ei) *Peter and Bill wants* a new car.

CG   Peter and Bill wants a new car.

CT   Peter and Bill wants a new car.

RD   Peter and Bill wants a new car.

PE   <Gram> The words "Peter and Bill" may be unclear. The structure of this sentence may be difficult to understand, or the sentence may be coordinated incorrectly. Should "Peter and Bill" agree with "wants?"

RW   Peter and Bill wants<<*_ Should "wants" be in its plural form? Replace "wants" by its plural form. *>> a new car.

GK   Peter and <"IF 'BILL' IS PART OF A COMPOUND SUBJECT (USED WITH 'AND' OR 'OR' AND ANOTHER NOUN), THEN IT SHOULD NOT BE USED WITH THE SINGULAR FORM OF THE VERB 'WANTS'.> Bill wants a new car.

LI   > Peter and Bill want a new car - if the subject has more than one item connected by <and> the verb must be plural.

(eii) *Peter or Bill want* a new car.

CG   Peter or [- The word 'Bill' does not agree with 'want'. -] Bill [- Consider 'wants' instead of 'want'. -] want a new car.

CT   > The word {Bill} does not agree with {want}. OR Consider [wants] instead of {want}.

RD   Peter or Bill want a new car.

PE   <Gram> The subject for "want" may be unclear. If it is "Bill," then "want" must agree in number with it. The structure of this sentence may need to be clarified.

RW   Peter or Bill want<<*_ Do the noun and the verb agree in number in Bill want? Change the appropriate one to make the noun and verb agree. *>> a new car.

GK   Peter or <THE SINGULAR NOUN 'BILL' MAY BE USED INCORRECTLY WITH THE PLURAL FORM OF THE VERB 'WANT'.> Bill want a new car.

LI   > Peter and Bill want a new car - if the subject has more than one item connected by <or> the verb must be singular.

(f) subject/verb agreement with subjects having post noun phrase

modification,

(fi) *The man with all the new computers are* very happy.

CG   The [- Consider 'men' instead of 'man'. -] man with all the new computers [- Consider 'is' instead of 'are'. -] are very happy.

CT   > Consider [men] instead of {man}. OR Consider [is] instead of {are}.

RD   The man with all the new computers are very happy.

PE   <Gram> The subject for "are" may be unclear. If it is "man," then "are" must agree in number with it. The structure of this sentence may need to be clarified.

RW   The man with<<*_ Does this need to be gender_specific? If not, replace "man with" with "person with." *>> all the new computers are very happy.

GK   The man with all the new computers <BE SURE YOU ARE USING 'ARE' WITH A PLURAL SUBJECT ('WE ARE', 'JILL AND DON ARE'). SINGULAR SUBJECTS JOINED BY 'OR' OR 'NOR' USUALLY TAKE A SINGULAR VERB ('JILL OR DON IS').> are very happy.

LI   > the man with all the new computers is very happy - the subject noun before the main verb and the verb must agree.

(fii) *The managers near the door doesn't* like work.

CG   The [- Consider 'manager' instead of 'managers'. -] managers near the door [- Consider 'don't' instead of 'doesn't'. -] doesn't like work.

CT   > Consider [manager] instead of {managers}. OR Consider [don't] instead of {doesn't}.

RD   The managers near the door doesn't like work.

PE   <Gram> The use of "managers" and "doesn't" may be unclear. If this is a subject-verb relationship, then "doesn't" must agree in number with "managers." If not, "managers" or "doesn't" may be part of a special phrase or may be unclear. Can you clarify?

RW   The managers near the door doesn't like work.

GK   The managers near the door doesn't like work.
LI   > the managers near the door don't like work - the subject
      noun before the auxiliary verb and the auxiliary verb
      must agree.

(fiii) *The man that could has falls were* tired.

CG   The man that could has falls [- This verb group may be in
      the passive voice. -] were tired.
CT   > This verb group may be in the passive voice.
RD   The man that could has falls were tired<<|STY:Passive|.
PE   <Gram> The subject for "were" may be unclear. If it is
      "man," then "were" must agree in number with it. The
      structure of this sentence may need to be clarified.
      <Gram> "Could" and "has" do not seem to belong
      together. Should one be removed? Has a word been left
      out? <Gram> If "that" refers to "man," it might need to
      be replaced by "who/whom." If not, the referent for
      "that" may be unclear. <Logc> "Could" expresses
      uncertainty and should be used only when this stance
      is appropriate.
RW   The man that could has falls were tired<<*_"were tired" is
      passive voice. Consider rephrasing the sentence to say
      who or what is doing the action. *>>.
GK   The man that <IF 'COULD' IS BEING USED AS AN
      AUXILIARY VERB, IT SHOULD BE USED WITH A
      SINGULAR PRESENT TENSE VERB (E.G., "COULD
      FIX").> could has falls < PASSIVE VOICE: 'WERE
      TIRED'. CONSIDER REVISING USING ACTIVE
      VOICE. SEE HELP FOR MORE INFORMATION.>
      were tired.
LI   > the man that could have fallen was tired - the main
      verb -- 'fall' 'complain' or 'go' -- must have the participle
      form 'fallen' or 'gone' in this context - the subject noun
      before the main verb and the verb must agree - the
      perfect verb following the auxiliary verb must have the
      form 'have' not 'has'.

(fiv) The old *books that was in the very new office near the manager's office was* expensive.

CG	The old [- Consider 'book' instead of 'books'. -] books that [- Consider 'were' instead of 'was'. -] was in the very new big office near the new manager's office [- Consider 'were' instead of 'was'. -] was expensive.
CT	> Consider [book] instead of {books}. OR Consider [were] instead of {was}.
RD	The old books that was in the very new big office near the new manager's office was expensive.
PE	<Gram> The use of "books" and "was" may be unclear. If this is a subject-verb relationship, then "was" must agree in number with "books." If not, "books" or "was" may be part of a special phrase or may be unclear. Can you clarify? <Usag> "New" is an adjective used in series. There may need to be a comma after it, or it may be part of a special phrase. See 'Tutorial' for a detailed explanation.
RW	The old books that was in the very new big office near the new manager's office was expensive.
GK	The old books that <BE SURE YOU ARE USING 'WAS' WITH A SINGULAR SUBJECT. ('IT WAS.').> was in the very new big office near the new manager's office was expensive.
LI	> the old books that were in the very new office near the manager's office were expensive - the subject noun before the main verb and the verb must agree - the verb in the relative clause and the noun must agree.

(g) possessive forms,

(gi) He *has take* the *dogs* bone.

CG	He has take the [- Consider 'dog's' or 'dogs'' instead of 'dogs'. -] dogs bone.
CT	> Consider [dog's] or [dogs'] instead of {dogs}.
RD	He has take the dogs bone.
PE	He has take the dogs bone.
RW	He has take the dogs bone.

GK   He <IF 'HAS' IS BEING USED AS AN AUXILIARY VERB, IT SHOULD BE USEDWITH A PAST PARTICIPLE VERB (E.G., 'HAS FIXED').> has take the dogs bone.

LI   >he has taken the dog's bone - the verb form after the auxiliary 'have' should have the past participle form - eaten, talked or sent - a noun in this context should be marked for possession - man's or mens'

(gii) The *secretarys computer* isn't working.

CG   The [-Consider 'secretaries' instead of 'secretarys'. -] secretarys computer isn't working.

CT   > Consider [secretaries] instead of {secretarys}.
     > Consider [secretary's] or [secretarys'] instead of {secretarys}.

RD   The secretarys<<|S{\bf PE:}~~:??|computer isn't working.

PE   The secretarys computer isn't working.

RW   The secretarys computer isn't working.

GK   The <SINGLE-WORD SPELLING ERROR.> <'THIS CONTEXT INDICATES THAT 'SECRETARYS' MAY NEED TO BE A POSSESSIVE (I.E., CHANGE TO "S' OR 'S")> secretarys computer isn't working.

LI   the secretary's computer isn't working - a noun in this context should be marked for possession - man's or mens'

(giii) *He send him book to that women.*

CG   He [-Consider 'sends' instead of 'send'. -] send him book to that women.

CT   > Consider [sends] instead of {send}.

RD   He send him book to that women<<|GRA:??|.

PE   <Prec> The use of words such as "they, each, them, he..." may cause the sentence to be vague. Could you be more specific?

RW   He send<<*_ Do the noun and the verb agree in number in He send? Change the appropriate one to make the noun and verb agree. *>> him book to that women<<*_

163

Is "that women" correct? This is unusually worded. Please study it to see if it will be clear to your reader. If not, reword it. *>>.

GK   <THE SINGULAR NOUN 'HE' MAY BE USED INCOR­RECTLY WITH THE PLURAL FORM OF THE VERB 'SEND'.> He send him book to that women.

LI   > he sends his book to those women - the pronoun before the noun needs the possessive form - a plural noun such as < women > needs a determiner like 'those' or 'these' not 'that' - the subject pronoun 'he' 'she' or 'it' and the verb must agree.

(h) pronouns in object positions,

(hi) The woman *is expects a important letters from he or she.*

CG   The woman is expects [- The word 'a' does not agree with 'letters'. -] a important [- Consider 'letter' instead of 'letters'. -] letters from [- Consider 'him' instead of 'he'. -] he or [- Consider 'her' instead of 'she'. -] she.

CT   > The word {a} does not agree with {letters}. OR Consider [letter] instead of {letters}. > Consider [an] instead of {a}. > Consider [him] instead of {he}.

RD   The woman is expects a important letters from he or she

PE   <Prec> The use of words such as "they, each, them, he …" may cause the sentence to be vague. Could you be more specific?

RW   The woman is expects<<*_ Is "is expects" correct? This is unusually worded. Please study it to see if it will be clear to your reader. If not, reword it. *>><<*_ Is this the correct form of the verb? Change the verb to make it match its subject in person and in number. *>> a important<<*_ Should this be "an" instead of "a?" Replace with "an." *>> letters from he<<*_ Is this the correct pronoun? Replace "he" with "him." *>> or she<<*_ Is this the correct pronoun? Replace "she" with "her." *>>.

GK   The woman <IF 'IS' IS BEING USED AS AN AUXILIARY VERB, IT SHOULD BE USED WITH A PAST PARTI-

CIPLE OR PRESENT PARTICIPLE VERB (E.G., 'IS FIXED' OR 'IS FIXING').> is expects <USUALLY 'A' IS USED WITH A SINGULAR NOUN OR NOUN PHRASE.> <USE 'AN' FOR WORDS BEGINNING WITH A VOWEL SOUND.> a important letters from <THE CONTEXT OF 'HE' INDICATES THAT YOU MAY NEED TO REPLACE IT WITH THE OBJECT PRONOUN 'HIM'.> he or she.

LI > the woman is expecting some important letters from him or her - a plural noun such as < letters > must have a determiner like 'some' not 'a' - the pronoun after the verb needs the accusative form --- him her or them - if the verb 'is' 'are' 'was' or 'were' is an auxiliary verb then the following main verb must be   in the present participle form --- 'being' agree.

(hii) The woman *wrote he* a long letter.

CG The woman wrote he a long letter.
CT The woman wrote he a long letter.
RD The woman wrote he a long letter.
PE <Usag> The personal pronoun "he" may be the wrong form of pronoun in this context. See 'Tutorial' for some better alternatives.
RW The woman wrote he<<*_ Is this the correct pronoun? Replace "he" with "him." *>>a<<*_ Is a comma missing? Consider inserting a comma as in "he, a" *>> long letter.
GK The woman wrote he a long letter.
LI the woman wrote him a long long letter - a pronoun in object position must have the form 'him', 'her' or 'them'.

(i) pronouns in embedded clauses,

(ii) *He writing* a book is unusual.

CG He writing a book is unusual.

CT   He writing a book is unusual.
RD   He writing a book is unusual.
PE   He writing a book is unusual.
RW   He writing a book is unusual.
CT   He writing a book is unusual.
LI   His writing a book is unusual - a pronoun in this position must have the form 'him', 'his', 'her', 'their' or 'them'.

(iii) The manager isn't happy *for they* to leave.

CG   The manager isn't happy for they to leave.
CT   The manager isn't happy for they to leave.
RD   The manager isn't happy for they to leave.
PE   <Usag> The personal pronoun "they" may be the wrong form of pronoun in this context. See 'Tutorial' for some better alternatives.
RW   The manager isn't happy for they to leave.
GK   The manager isn't happy for <THE CONTEXT OF 'THEY' INDICATES THAT YOU MAY NEED TO REPLACE IT WITH THE OBJECT PRONOUN 'THEM'.> they to leave.
LI   the manager isn't happy for them to leave. - a pronoun in this position must have the form 'him', 'her' or 'them'.

(j) modal auxiliaries

(ji) The new secretary *cans type* very fast.

CG   The new secretary cans type very fast.
CT   The new secretary cans type very fast.
RD   The new secretary cans type very fast.
PE   <Reln> The use of "cans type" may not be correct. Make sure that "cans" can perform the action named by "type". If not, then you may need to clarify this sentence.
RW   The new secretary cans type very fast.
GK   The new secretary cans type very fast.
LI   > the new secretary can type very fast - a modal verb such as 'can' or 'must' does not agree with the pronoun or verb.

(jii) The new manager *type must* all the letters.

CG   The new manager type must all the letters.
CT   The new manager type must all the letters.
RD   The new manager type must all<< I GRA : ?? I the letters.
PE   The new manager type must all the letters.
RW   The new manager type must all the letters.
GK   The new manager type must all the letters.
LI   the new manager must type all the letters - the word
     <must> which is a(n) <modal> has been moved from
     sentence position <5> to <4> - the verb form following
     a modal like 'can' or 'should' needs the base form -
     'write' or 'eat'.

(k) forms of auxiliaries and main verbs,

(ki) The secretary *could has go* to the bank.

CG   The secretary could [- Consider 'have' instead of 'has'. -]
     has go to the bank.
CT   > Consider [have] instead of {has}.
RD   The secretary could has go to the bank.
PE   <Gram> "Could" and "has" do not seem to belong to-
     gether. Should one be removed? Has a word been left
     out? <Logc> "Could" expresses uncertainty and should
     be used when this stance is appropriate
RW   The secretary could has go to the bank.
GK   The secretary <IF 'COULD' IS BEING USED AS AN
     AUXILIARY VERB, IT SHOULD BE USED WITH A
     SINGULAR PRESENT TENSE VERB (E.G., "COULD
     FIX").> could <IF 'HAS' IS BEING USED AS AN
     AUXILIARY VERB, IT SHOULD BE USED WITH A
     PAST PARTICIPLE VERB (E.G., 'HAS FIXED').> has
     go to the bank.
LI   > the secretary could have gone to the bank - the perfect
     verb following the auxiliary verb must have the form
     'have' not 'has' - the main verb -- 'fall' 'complain' or 'go'
     -- must have the participle form 'been' or 'fallen' or
     'gone' in this context.

(kii) If they weren't in the office the men *couldn't have see* the secretaries.

    CG   If they weren't in the office the men couldn't have see the secretaries.

    CT   If they weren't in the office the men couldn't have see the secretaries.

    RD   If they weren't in the office the men couldn't have see the secretaries.

    PE   <Punc> There may be a structural problem in this sentence. The words around "couldn't have" may be the source of the problem. Is a comma needed at some point?

    RW   If they weren't in the office the men couldn't have see<<*_ Is "have see" correct? This is unusually   worded. Please study it to see if it will be clear to your reader. If not, reword it. *>> the secretaries.

    GK   If they weren't in the office the men <WHEN 'COULDN'T' IS USED AS AN AUXILIARY VERB OR MODAL, AND IS FOLLOWED BY 'HAVE', THE FOLLOWING WORD SHOULD BE THE PAST PARTICIPLE FORM OF THE VERB 'SEE'.> couldn't <IF 'HAVE' IS BEING USED AS AN AUXILIARY VERB, IT SHOULD BE USED WITH A PAST PARTICIPLE VERB (E.G., 'HAVE FIXED').> have see the secretaries.

    LI   > if they weren't in the office the men couldn't have seen the secretaries - the main verb – 'write' 'eat' or 'walk' - - must have the past participle form 'written' or 'walked' in this context.

(l) adjectival and verb complementation.

(li) They were *happy to seeing* them.

    CG   They were happy to seeing them.

    CT   They were happy to seeing them.

RD   They were happy to seeing them.
PE   They were happy to seeing them.
RW   They were happy to seeing them.
GK   They were happy to seeing them.
LI   they were happy to see them - the pronoun in this clause needs the accusative form - 'him' 'her' or 'them' - the verb after the infinitive 'to' must have the base form - write, eat or sleep.

(lii) *He were* happy *for they to leaves.*

CG   He [- Consider 'was' instead of 'were'. -] were happy for they to leaves.
CT   > Consider [was] instead of {were}.
RD   He were happy for they to leaves.
PE   <Gram> The subject for "were" may be unclear. It it is "he," then "were" must agree in number with it. The structure of this sentence may need to be clarified. <Usag> The personal pronoun "they" may be the wrong form of pronoun in this context. See 'Tutorial' for some better alternatives.
RW   He were<<*_ Do the noun and the verb agree in number in He were? Change the appropriate one to make the noun and verb agree. *>> happy for they to leaves.
GK   He <'WERE' IS USED WITH A PLURAL SUBJECT ('WE WERE`), EXCEPT IN THE SUBJUNCTIVE ('IF HE WERE'). SINGULAR SUBJECTS JOINED BY 'OR' OR 'NOR' USUALLY TAKE A SINGULAR VERB ('EITHER JILL OR DON WAS').> were happy for <THE CONTEXT OF 'THEY' INDICATES THAT YOU MAY NEED TO REPLACE IT WITH THE OBJECT PRONOUN 'THEM'.> they to leaves.
LI   > he was happy for them to leave - the pronoun in an embedded clause needs to be of the form --- them her or him - the subject pronoun 'he' 'she' or 'it' and the verb must agree.

(liii) All the customers *complained the manager.*

CG   All the customers complained the manager.

169

CT   All the customers complained the manager.
RD   All the customers complained the manager.
PE   All the customers complained the manager.
RW   All the customers complained the manager.
GK   All the customers complained the manager.
LI   > all the customers complained prep the manager - A
        < prep > has been inserted at sentence position < 5 > .

General observations

A number of observations can be made about the relative performance of the seven programs and a number have been made about the older programs (Obermier, 1990; Williams, 1991)[16] - some positive but with the more linguistically informed comment being negative (Bowyer, 1989). In addition, much could be said about the programs' treatment of individual sentences, the presentation of advice and the wording of such advice. The criteria for such observations, for both programs and individual sentences, include accuracy of error detection, value of suggestion or repair strategy, range of language and problems covered, speed of operation and general ease of use. The trade-off between these critrea is doubtless to some degree a matter of personal choice. However, accuracy of error detection, value of suggestion/repair strategy combined with range of language and problems covered are of primary importance - speed and ease of use are not of much value if they do not identify errors.

On syntactic as opposed to morphosyntactic problems, that is when words not only have the wrong form but are also in the wrong place - the sentence as a whole has errors of arrangement - then LINGER comes off better. This is to a large extent understandable as it performs a complete sentence analysis as opposed to the other programs that work with bits of a sentence rather than the whole. Most of the commercial programs have a tendency to work 'outwards' from certain more easily identifiable items, categories and constituents - see later - and consequently underperform on the sentence as a whole.

For speed and ease of use the commercial programs *Correct Grammar for Windows, Right Writer for Windows, Grammatik for Windows, Reader, PowerEdit, CorrecText* and LINGER are the clear leaders. With little to chose between the first five for speed - *PowerEdit* is slightly slower - they regrettably miss a number of errors that the less developed and consid-

[16] Reviewer's Notebook in *Byte*. September 1991 pp310-312.

Table 1. Example statements
(+ = correct error detection and an accurate response/advice; ? = either the error detection or the response is questionable; 0 = the grammatical error was not detected.)

Sentence	Program						
	CG	CT	RD	PE	RW	GK	LI
He writes *an letter*.	+	+	+	+	+	+	+
He writes *an letter new*.	+0	+0	00	+0	+0	+0	++
This books is very old.	++	++	00	++	++	00	++
I want to send *these very interesting book* to him.	+	+	0	0	0	+	+
The bear eat *that berries old*	?	?	?	?	?	?	?
He has *many evidence*.	+	+	0	+	0	+	+
There is *much reasons*.	0+	0+	00	+0	00	+0	++
An dogs big was in *those house*.	++0+	++0+	0000	0000	000+	0++0	++++
The man expect the letter.	+	+	0	+	+	+	+
They is happy.	+	+	0	+	+	+	+
An young dog arent eats the smelly fish	+00	+00	+00	++?	+00	+0+	+++
The dog has eat the bone so it *cant being* hungry.	+00	+00	000	000	000	++0	+++
The dog didnt eating another fishes.	+00	+00	000	000	000	0+0	+++
Although *they was* very tired they *persuaded the manager to leaves*.	00	+0	00	+0	+0	+0	++
A good book *for she to reading are* under the desk.	0+	0+	00	00	00	+0	++?
Peter and Bill wants a new car.	0	0	0	+	+	+	+
Peter or Bill want a new car.	?	?	0	+	0	0	+
The man with all the new computers *are* very happy.	+	+	0	+	0	+	+
The managers near the door doesn't like work+	+	0	+	0	0	+	
The *man that could has falls were* tired.	00	00	00	++	00	?0	++
The *old books that was in the very new office near the manager's office was expensive*.	++	++	00	?0	00	+0	++
He *has take the dogs bone*.	0+	0+	00	00	00	+0	++
The *secretarys computer* isn't working.	+	+	+	0	0	+	+
He *send him book to that woman*.	+00	+00	?	000	+0+	+00	+++
The woman is expects a important letter from he or she.	0++0	0++0	0000	0000	+00+	+0+0	++++
The woman *wrote he* a long letter.	0	0	0	+	+	0	+
He writing a book is unusual.	0	0	0	0	0	0	+
The manager isn't *happy for they to leave*.	0	0	0	+	0	+	+
The new secretary *cans type* very fast.	0	0	0	+	0	0	+
The new manager *type must* all the letters.	0	0	+	0	0	0	+
The secretary *could has go* to the bank.	0	0	0	0	0	0	+
If they weren't in the office the men *couldn't have see* the secretaries.	0	0	0	?	+	+	+
They were *happy to seeing* them.	0	0	0	0	0	0	?
He *were* happy *for they to leaves*.	+00	+00	000	0+0	+00	++0	+++
All the customers *complained the manager*.	0	0	0	0	0	0	+

erably slower LINGER picks up. LINGER, by contrast, will attempt to tell the user when something is wrong and has a facility for learning a new word before a parse is attempted. However, if it does not have a rule to cover the construction then it will not be able to give relevant advice, and as is evident, on such occasions can produce non relevant advice. A summary of error detection of a grammatical nature is presented in Table I.

In general, on the evidence presented here, it can be concluded that for the typical learner of English, the programs as a group under-perform, although they underperfom to varying degrees and in varying ways. Words and word patterns that are characterised by more easily identifiable morphosyntactic features and the relative closeness of related words and general size of word patterns fare better, for example close subject-verb or article/determiner-noun patterns. This is thoroughly understandable and explains why, for the kinds of words and word patterns that are susceptible to this treatment figure more highly in the table above. Such performance might suggest that a full analysis of each sentence is being performed. It is when, however, the irregular sentence patterns are considered, usually the second or later sentence(s) in each grouping that the commercial programs begin either to mis-identify structural problems or miss advise on relevant matters, or simply miss. The reasons for this, the major one being the fact that no complete sentence analysis seems to be performed, are addressed next.

The program architecture which determines performance - transparent,semi-transparent, semi-opaque and opaque systems

In addition to the general observations noted at the end of the previous section, a principal observation about these programs is that we are not in any way comparing like with like. The six commercial programs are 'finished' programs, in that they 'work' at speed, work with word processors and are apparently actually used by people. The downside is that they do not really work to any satisfaction in the sense of picking up on a number of the errors that we know to exist in certain sentences. None of the commercial programs seems to cope with words that are in the wrong place in a principled fashion. It is fairly easy to trip them up, as has been demonstrated. Conversely, it is easy to demonstrate that on certain counts LINGER does, from a user's perspective, a better, if slower job in a much restricted domain, but LINGER's coping is very limited and it would take very little variation in sentence and especially lexical input to reduce the program to irrelevance. There is a general underperformance by all the programs.

From a more critical perspective, as important as that the programs do or do not give certain performances is how they do what they do and conversely why they do not do what they do not do. If in addition to being able to see what they do we are able to see how they do it - if the program's resources and processes are transparent - and, further, if these resources and processes are manipulable, then so much the better. This is not to suggest that our typical user would be able to benefit directly from such openness but that tutors might reasonably be tempted to attempt manipulation, or at the very least, know the limitations, and the design and structural reasons for these limitations, of the system. By such manipulation as is possible we may be able to determine if performance is constrained by design, architecture or implementation and if weaknesses that can be identified in any of these can be modified to effect a more acceptable performance.

The principal cause of underperformance by the commercial programs is that they do not treat the sentence, in terms of the words entered by the user, in a holistic fashion. Although words are given grammatical class categories, sometimes correctly and sometimes incorrectly through commission or omission, these are not sufficiently linguistically contextualised to provide unique and complete syntactic structure(s). Either a 'structure' is produced which vacilitates between different word class assignments - *RightWriter* and *Grammatik* - some patently implausible, or, in other cases where only one word class assignment has been made, the actual or reconstructed tree is at odds with what most grammars of English would allow as well-formed sentences.

Program requirements

It was suggested above that a minimally acceptable grammar checking program needs at least three elements. Firstly a large dictionary against which to match incoming words: match in this context means, firstly recognising them as correctly spelled and secondly, assigning to each the grammatical category(ies) to which it belongs together with any other syntactic and/or morphosysntactic information which the program uses in constructing sentences (and importantly, the intermediate structures of the phrase and the clause). Secondly, a process for putting those words into acceptable combinations (and some way of representing acceptable combinations) is required, noting a distinction for rule-based systems between the rules themselves and the operation of these rules. Thirdly, a process for recognizsing and reacting to sequences and forms of words for which no acceptable sequence or arrangement can be found, is necessary.

A program that has these three elements and allows access to these elements and their operation is considered for our purposes as a transparent program whereas one which allows access to none of these is considered an opaque program. To these three, a further component might be added and that is the extent to which the resources - information and structures and the processes used by the program in determining sentence structure, differentiatng between well-formed and non- well-formed sentences and describing the deficiencies in the latter - are used and made available to the user in the presentation of the results of the program's work and to what extent these are accessable and manipulable. Using this framework, the seven programs examined can be divided into opaque, semi-opaque, semi-transparent and transparent.

Properties of an adequate grammar checking program

An adequate program, apart from performing in a way to identify all grammatical errors and giving relevant information about the error, access and manipulation can be considered from at least the four following perspectives:

(a) the possibility of seeing what the relevant word and rule 'dictionaries' contain, what items they hold and what the structure(s) of these items are,

(b) the possibility of modification of the word and rule dictionaries, that is either adding new items or altering existing items,

(c) gaining knowledge of the operation of the various grammatical rules, and,

(d) determining from what is presented to the user how the program has used its various resources to perform its computation - what category(ies) words have been identified as and what patterns these have been put into, and how these patterns reflect - and can be seen to reflect - the program's representation of what is an acceptable sentence.

Opaque programs

Correct Grammar, *Reader* and *CorrecText* are similar in not providing any way into their resources or sentence representation processes. They do not allow access to their word or rule dictionaries in order to add more words or rules or gain further information about words already in there. Neither is it possible to see the nature of the processes that guide the representation of a sentence. Nor is it possible to see how they determine how a sentence has gone wrong, rather, by putting sentences through the programs it is merely possible to see what the programs do when they detect a problem. We cannot access their representation of the sentence - either well or ill-formed. Rules can be turned 'on' or 'off';

CorrecText offers the user the possibility of using the following rules groups:

Ungrammatical Expressions	No
Noun Phrase Consistency	No
Mass vs Count; A vs An	No
Verbal Group Consistency	No
Subject-Verb Agreement	No
Pronoun Errors	No
Clause Errors	No

where the 'no' can be changed to a 'yes' to make the rule operative, while under 'Style' Reader lists the following:

Contraction	No
And/or/but at start	No
Bad use of more/less	No
Passive voice	No

Preposition at end	No
Split infinitive	No
Sentence has no verb	No

which likewise can be activated or not.

Correct Grammar for Windows allows not only some rules to be turned or off but also allows for slight modifications made to, for example the number of words allowable in a split infinitive - see Fig 1 which shows the lowest level of access permitted to various features;

Fig 1
Correct Grammar for Windows Lowest Level of Access

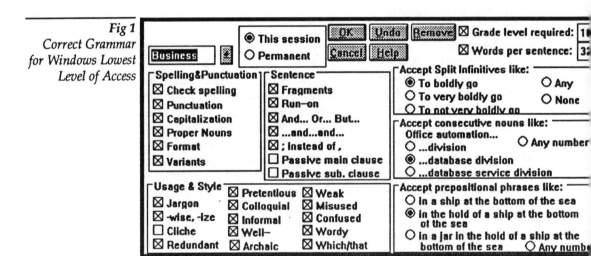

The actual value of the rules themselves and the opportunity to use or not use them, or to fine tune them within their limits is, from a strictly grammatical perspective, debatable. None of the programs gives full explanation of what actually comprises the rules. *CorrecText* offers examples of the following kind:

Verbal Group Consistency

The verbs in your verb group should all be in the same tense. You can usually correct this by changing the tense.

Problem: We *could have drove* to New York in August.

Solution: Use *driven* instead of *drove*, OR, remove *could have*. "[17]

Presentation

Presentation to the user by *Reader* is minimal, simply a marked up copy of the original file - only a 'batch' file mode of processing is offered. In addition, reader appears to have only one catch-all grammatical error marker, which detected less than 10% of the known errors. Both *CorrecText* and *Correct Grammar for Windows* permit the user to see what errors are being detected and how they are being diagnosed - they have both 'batch' file and interactive modes - with the option in the latter of correction at that stage - an example from the latter program is presented in Fig 2.

[17] *The Complete Writer's Toolkit* manual. p107.

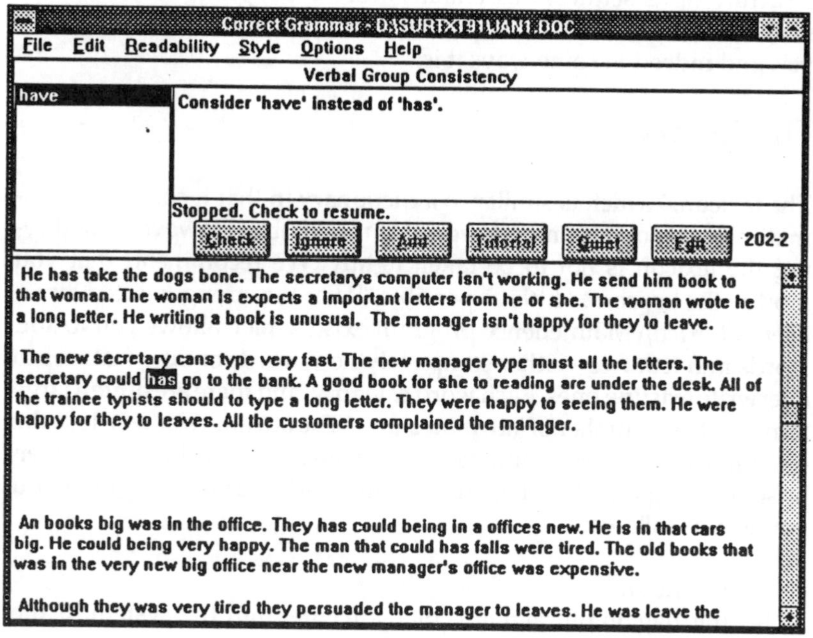

Fig 2
Correct Grammar
for Windows error
detection and
diagnosis.

Use of resources

With respect to how the programs use their resources to inform their output and to inform their users about the output, all three of these programs do not fare very well. *Reader* provides almost nothing of value from the perspective of the kinds of errors in these texts and the level of user that is making them. *CorrecText* and *Correct Grammar* provide an identical level of correction for a number of the errors encountered - about one third - and much of it is minimally relevant. They do not, however, show the user the structure of the sentence that has been produced so that they do not really exploit their own resources. As a result, all of these programs can be considered opaque.

Semi-opaque programs

PowerEdit is a very large program that, while not permitting very direct access to its resources in general, gives copious information about the structure of its settings - essentially word type/group sizes. All this suggests a comprehensive approach to the problems of sentence analysis and indeed in some ways it is.

Dictionary

The lexicon, "which is similar to a dictionary in that it contains grammatical and semantic information"[18], is modifiable. However, "modifying the lexicon is not to be taken lightly. We recommend that you modify the lexicon as little as possible."[19] For the strong-hearted, however, a Lexicon Maintenence program exists which allows you to add words and features - either completely new words or new features to overwrite existing ones. Each word category, and there are quite a large number, has a number of 'attributes' and "The attributes of words in the lexicon define the word and its possible usage". [20] Attributes are given by setting 'flags' for the relevant word, the guidance for setting up plural nouns is as follows;

[18] *PowerEdit User's Guide* pD-1.
[19] *PowerEdit User's Guide* pD-1.
[20] *PowerEdit User's Guide* pD-5

"plural noun
Set this flag on plural nouns that you enter into the system. If the noun that you are entering into the system is a regular noun that

inflects with "-s" or "-es" as its plural, then you do not need to enter its plural form. If the noun is irregular, or ends with "-ings" then you do need to enter the plural form of the noun into the system. Examples of irregular nouns are: foot/feet, leaf/leaves, life/lives, and ox/oxen.

If the noun ends in "y" and takes "-ies" as its plural, it also needs to be entered. If it is a word that has the same form for singular and plural (fish), it needs to be entered into the system with the plural noun flag set and the singular nounflag set. A noun that can also be an adjective cannot have the plural noun set alone. It also needs to have the singular flag set as well. Examples of these kinds of nouns are: Swedish, Finnish, and Polish. When you set both the plural noun flag and the singular noun flag, the system will be able to set the proper determiners, articles, and adjectives necessary for number agreement.[21]

This range of features which attaches to the various word categories is impressive in itself and suggests an understanding of the complexity of natural language which most of the other programs do not.

Grammar

There is no overt grammar in the sense of a set of accessible or viewable rules about how sentences as a whole are defined. Rather, the user is given a long list of 'maximums' which singularly and conjoinedly define the permissible shape of a sentence. These have the default values noted in Fig. 3 but can all be changed. Again, the number of items included in this list suggests an appreciation of the sutbtly of natural language.

"Max. Pronouns If the number of pronouns multiplied by the setting exceeds the words in the sentence, a message appears.

Default: 7   Message: 248"

Example:
He send him book to that women. (giii)

[21] *PowerEdit User's Guide* pD-11.

<Prec> The use of words such as "they, each, them, he . . ." may cause the sentence to be vague. Could you be more specific?

"Result:
This sentence generates message number 248 because the number of pronouns ( ) x setting (7) exceeds the number of words in the sentence."[22]

[22] *PowerEdit User's Guide* pVI-26.

Fig 3
PowerEdit list of default values.

Maximun Sentence Length	38	Max. Nominalizations 2	5
Max. Modifiers	4	Max. Modifiers 2	50
Max. Adjectives	5	Max. Explan. Elements	4
Max. Explanatory Words	9	Max. Verb Phrase Words	6
Max. Descriptive Words	11	Max. Adjectives in a Row	4
Max Punctuation	5	Max. Nouns in a Row	5
Max. Nominalizations 1	3	Max. Noun Phrase	6
Max. Introductory Words	4	Max. Verbs in Phrase	4
Max. Choppy Elements	0	Max. Interruption	5
Max. Rel. Clauses in Row	3		
Max. Subordinate Clauses	8		
Max. Adverbs in a Row	5		
Max. Preposition Phrases	10		
Max. Nouns	4		
Max. Pronouns	7		

Presentation

PowerEdit gives the user on screen feedback of a detailed nature using the relevant messages as presented earlier a success rate of approximatelt 50%. Relevant problem areas are highlighted and the programs gives the user a range of possibilities relating to seeing more information about the error type and editing the problem area. Not presented, however, is information about word class categories, nor, more importantly, how patterns of such categories are structured in accordance with patterns or rules that set the well-formed sentence off from the ill-formed one.

Use of resources

The user intending to work out exactly what is a permissible sentence structure and what is not is therefore required to put a number of tightly controlled sentences through the program because, although as with the wholly opaque programs some turning on or off or setting of certain quantities of words and word patterns is permitted. As a result it might be termed a semi-opaque program.

Semi-transparent programs

RightWriter is the one program that notes the second language learner as a possible user. It is a claim that should not be taken at face value.

Dictionary

RightWriter permits no access to its dictionary. Indeed, the word 'dictionary' does not appear even in the index of the accompanying documentation. There is no way in which the user can see what words have what features/attributes and what those attributes might be. Similarly, there is no obvious way in which the user can change any of the settings which must exist for it to identify the errors that it does.

Rules

Access to and iformation about 'rules' is limited in the same way as, for example *Reader* and *Correct Grammar*, in being confined to turning such rules on or off. In this sense it is more limited than *Correct Grammar* in that some of these are not flexible - being either on or off, nor are examples given, and, as a set, they are vastly more impoverished than, for example, *PowerEdit's*. The relevant rules are classed as either Grammar or Style rules and, in addition, there is a feature called Sentence Structure Recommendations. These are shown in Fig. 4.

Fig 4
Grammar Rules,
Style Rules and
Sentence Structure
Recommendations in
RightWriter.

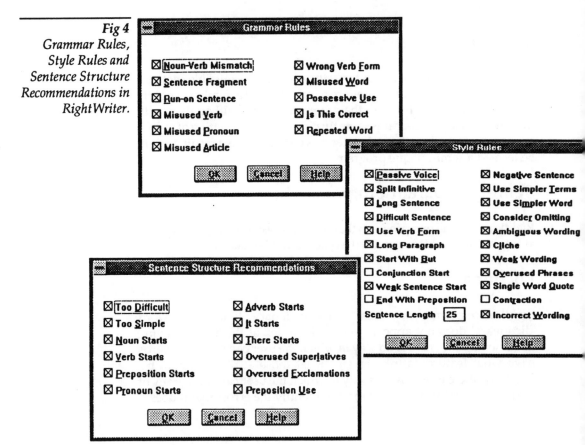

Presentation

Interactive or batch file processing is possible and the former is shown in Fig. 5.

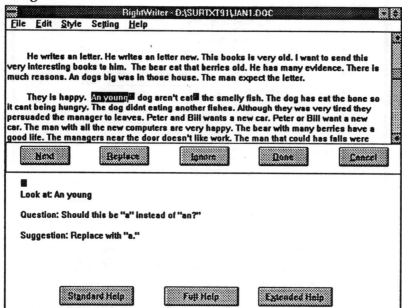

*Fig 5
Interactive file
processing in
RightWriter*

Use of resources

RightWriter does score over all the other programs considered so far, however, in the respect that it presents its analysis to the user in terms of sentence analysis and representation - *RightWriter* gives the user a parse tree of sentences in which it has found errors.

The value of the information given in any such parse tree naturally depends on how well the individual words have been initially categorised from the dictionary and how well, given the facts of multiple class membership, a particular occurence in a particular context has been uniquely categorised. As Fig. 6 shows, in both cases there are not only question marks over the initial multiple assignments for "all" and "to" but the indeterminacy for these plus "leaves" has vitiated the resulting sentence 'structure' and crippled the consequent error detection.

Fig 6
RightWriter
sentence parse trees.

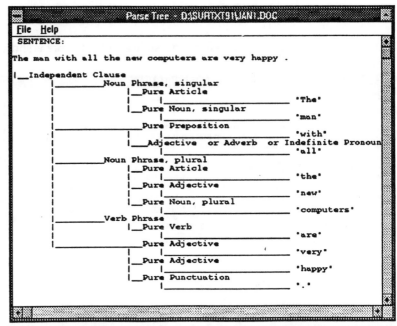

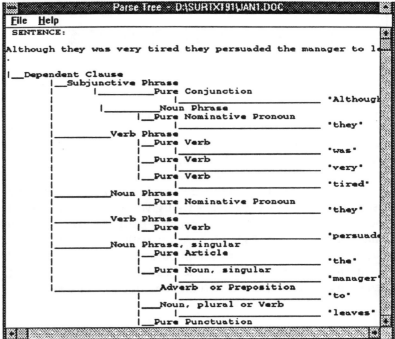

While it is impossible for the user to get at either the dictionary or the rules that are clearly being used to some purpose, the tree structure information lets the user view a form of result. The value of this representation can really only be judged by an individual user but in general, given its general indeterminacy over assigning a unique word class category to each word, and as a result vacillating over the final sentence structure and as a further consequence not identifying errors, it would seem of little practical value to the serious user. *RightWriter* might well be considered a semi-opaque program.

A semi-transparent program - Grammatik

Grammatik is one of the oldest grammar (and style) checking programs and offers, potentially, the greatest access to its resources and use of resources.

Dictionary

Access to and manipulation of *Grammatik's* dictionary is not especially easy in that there is no easy way to see what is in the dictionary. It is possible to add and delete words and through such a combination to effect replacement of currently held words. However, this is a somewhat inefficient process and rather hit and miss. As a result, we get to know what is in the dictionary in terms of word class (part of speech) assignment by looking at what the program gives as analyses.

Grammar

As with the previous programs, there is a distinction between rules which form the core of the program - subject-verb agreement, the use of 'a' and 'an' - that the user has no access to, and there is a selection of rules that the user can either use or disable. For Grammatik, the latter are shown in Fig. 7.

Fig 7
Optional rules
available in
Grammatik.

RULE CLASSES

◉ Grammatical ○ Mechanical ○ Style ○ User-Defined

Grammatical Rule Class

☒ Relative pronoun   ☒ Adverb
☒ Infinitive   ☒ Number agreement
☒ Incomplete sentence   ☒ Verb agreement
☒ Possessive form   ☒ Article
☒ Homonyms   ☒ Comparative
☒ Pronoun   ☒ Preposition
☒ Double negative   ☒ Incorrect verb form

[?]   [OK]   [Cancel]

Grammatik provides a limited form of access to dictionaries and some of its rules. In addition, it provides the resources with which to create more rules or modify existing ones as in Fig. 8.

Fig 8
Rule Editor in
Grammatik.

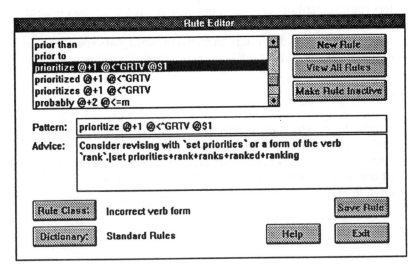

Rule Editor

prior than
prior to
prioritize @+1 @<^GRTV @$1
prioritized @+1 @<^GRTV
prioritizes @+1 @<^GRTV
probably @+2 @<=m

[New Rule]
[View All Rules]
[Make Rule Inactive]

Pattern: prioritize @+1 @<^GRTV @$1

Advice: Consider revising with `set priorities` or a form of the verb `rank`.|set priorities+rank+ranks+ranked+ranking

[Rule Class:] Incorrect verb form   [Save Rule]

[Dictionary:] Standard Rules   [Help]   [Exit]

The principal parts of this rule are: the @ indicates that this is a parsing rule, the +1 that the rule picks up the 'error' in front of the offending word, and the '^' in @<^GRTV indicates that the rule is to apply to words that form the subset intersection' of the following part of speech classes - G = present participle, R = past participle, T = past tense of verb and V = base or infinitive form of verb.[23] The result of this rule is that where 'prioritising/prioritised' is found in the text the advice is offered to the user. Users can add as many of this type of rule as imagination or ingenuity permit.

The rules are not so much rules in the conventional linguistic sense but rather actions to be performed on specified linear sequences of words. There is no structure in the sense of phrase or clause boundaries - no 'edges' are defined or used. This is very evident when the number of words that intervene between an identified word and the word or words that it relates to. There is a 'span' rule which allows the user to specify that the number of words between say a subject and verb or determiner and noun can be up to nine words. Unfortunately, that stretch of nine words can include a number of phrases and clauses. In

"The old books that was in the very new office near the manager's office was expensive."

Grammatik identified the first subject-verb mismatch but failed to identify the second - the 'span' between the relevant items being 11 words but, linguistically mor relevant, those 11 words included a relative clause which in turn included a subject complement with a prepositional phrase. Such a non-linguistically motivated dimension is likely to involve a number of linguistic items and, therefore, to be of little practical use.

Most of the accessible rules are of this second type, make up a very large proportion of the total 'rules', are not especially helpful or relevant from a strictly grammatical perspective. Unfortunately, as will be seen in the following section, a number of the more relevant parsing rules are inaccessible.

[23] *Grammatik for Windows User's Guide* pp152-153.

Presentation

The nature of the process whereby *Grammatik's* representation of a sentence is achieved is not directly available to the user. Below (Fig. 9) are examples of what the user can see of the results of a sentence representation but it is not clear how this representation is achieved. The manual tells us that artificial Intelligence techniques - "such as rules, backtracking and heuristics" [24] - are employed together with linguistic information but it is difficult to determine what techniques or linguistic information contributed to the assignment of the parts of speech in a sentence [25] and how these are then determined to be legal sequences in English. *Grammatik's* rules as such are really processes that result in these observable sequences of parts of speech representations but how they are arrived at is unclear.

[24] *Grammatik for Windows User's Guide* p77.
[25] *Grammatik for Windows User's Guide* p77.

Fig 9
Sentence representation in Grammatik.

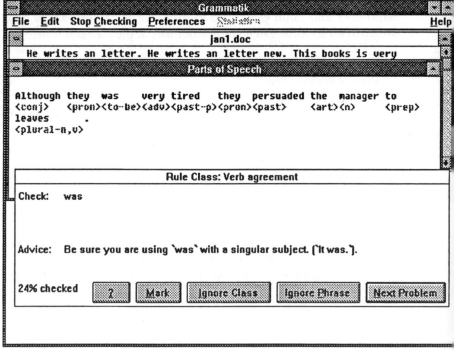

Unfortunately, in sentences that are regarded by the program as correct the parts of speech analysis not available (it can be forced by adding an extra full-stop). This is a major limitation to learning correct patterns and it would seem to be somewhat perverse to give an analysis, often a wrong analysis, of incorrect sentences. It also makes it that much more difficult to determine what is accepted as correct analyses. The principal way of determining part of speech assignment and therefore sentence, as opposed to phrasal or clausal constituency is by a process of trial and error, firstly by determining part of speech by passing both correct and incorrect sentences through the program and carefully noting what works where and what doesn't and why.

Use of resources

As with *RightWriter*, being able to see the structure assigned to a sentence could help the informed user see where things have gone wrong - both in the original sentence and in the program's analysis. However, as with *RightWriter*, the problem of initial word class assignment and the subsequent choosing between alternative assignments in particular contexts would seem to betray the fact that a full sentence analysis using phrases and clauses as objects with properly defined 'edges' is not being performed. Again, the success rate in error identification - about a third - would seem to confirm this While not totally so therefore *Grammatik* can be considered semi-transparent, although after attempts to manipulate the program elements the user may be inclined to describe it as semi-transparent.

189

LINGER a relatively transparent program

LINGER is an expert systems approach to grammatical error checking and comprises a dictionary, a rule base, an inference engine, a set of checking conditions and two levels of output to the user which draw on the previous resources and processes. LINGER first assigns categories to words and then builds combinations of word classes in accordance with conventional rewrite rules (embodying an analysis based very closely on Quirk and Greenbaum (1990) to ensure that the correct word is in the correct position and then applies further context sensitive rules to ensure that these words have the correct form.

Dictionary

The dictionary consists of a list of words, each entry giving the surface form of the word, its grammatical category or categories and other relevant features with their attendant values; e.g.

word([t,h,e,y],[cat(pnoun),type(persal),person(3),plurality(p)],[they]).
word([h,e],[cat(pnoun),type(persal),person(3),plurality(s)],[he]).
word([t,r,a,i,n,e,e],[cat(adjective),vowel(n)],[trainee]).
word([i,m,p,o,r,t,a,n,t],[cat(adjective),vowel(y)],[important]).
word([a,m],[cat(verbcop),person(1),plurality(s),tense(pres)],[be]).
word([i,s],[cat(verbcop),person(3),plurality(s),tense(pres)],[be]).

against which input is checked. 'cat' indicates the word's grammatical category (ies), while the relevant morphosyntactic features are given their respective values.

Rules

Partly context free rules provide the principal structures for grammatical rules which are used to build a representation of a sentence.

mcl ==> [sphr,vphrcop,advcomp].
mcl ==> [sphr,vphrcop,scphr].

vphrcop ==> [verbcop].

vphrcop ==> [verbcopn].
vphrcop ==> [verbcop,neg].
vphrcop ==> [axphr,verbcop].
vphrcop ==> [axphr,inter,verbcop].

The rules are used by a chart parser which tempers the conventional prolog process of exhaustive depth-first search. This whole sentence approach produces a complete parse tree for most sentences for which the structures (rules) exist, rearranging if necessary by movement, insertion and/or deletion of word categories, the original words.

Presentation

The first level of output to the user is a tree of the form in Fig. 10.

```
██                              user                              ██
 sentence
  mcl
   sphr
    np
     detp
      det >>>>>>>:>  the
     nounh
      noun >>>>>>>>:>  secretary
   vphrins
    axphr
     modaux >>>>>>:>  could
    inter
     perf >>>>>>:>  has
    verbins >>>>:>  go
   advcomp
    prepp
     prep >>>>>>:>  to
     np
      detp
       det >>>>>>>>>:>  the
      nounh
       noun >>>>>>>>>:>  bank

       ********* Press return to continue. *********
```

Fig 10
First level output in LINGER.

Context sensitive rules are then applied to the structures produced by the rewrite rules which are similar in their operation to the rules used by *Grammatik* - but not in form or content - in that when certain context determined sequences are encountered relevant actions are performed and advice offered, resulting in the second stage of user information.

```
prep >>>>>>:>  to
np
  detp
    det >>>>>>>>>>:>  the
  nounh
    noun >>>>>>>>>:>  bank


  ********  Press return to continue.  ********

===============================================================
Your original sentence was:-
> the  secretary  could  has  go  to  the  bank
The correct version of the sentence should be:-
> the  secretary  could  have  gone  to  the  bank
===============================================================
the perfect verb following the auxiliary verb must have the form 'have' not 'has'

the main verb – 'be' 'seem' or 'appear' – must have the participle form 'been' or 'appeared' i

Do you want another go [y]es ... or [n]o.
```

Use of resources

LINGER allows full access to its dictionary, rules and checking processes. The downside is that in order to make any modifications the user has to be well acquainted with the program and its workings, especially the way in which it links dictionary entries, rules for combining words and the structure of the corrections it makes (below).

```
in vphrins
if [ exists_str(inter,perf),
    not_has_features(perf,[person(1),plurality(p),tense(X)]) ]
then [change_word(has,have),
comment([the,perfect,verb,following,the,auxiliary,verb,must,have,the,
form,"'have'",not,"'has'"]) ].
```

The purpose of which rule is to ensure that when the word 'has' is used in the perfective form in a verb group of the type 'could (perf) main verb' it appears as 'have' not as 'has'.

As such it is highly transparent (to the initiated and prolog programmer) but not especially accessible to the casual user and certainly not to our typical end user. It fully uses its resources in presenting to the user how it has categorized each word in context, the sentence (and phrases and clauses) it has built and the points where it has detected errors. In addition, however, to its virtue in this aspect, it has only a limited coverage and where it runs out of information, generally because the input is too badly structured, it comes up with very erratic analyses, as in Fig. 12.

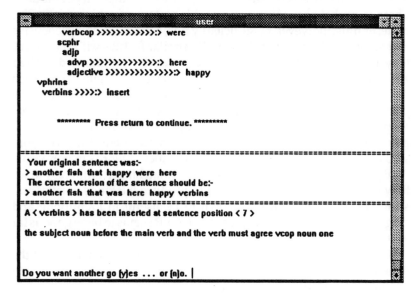

Fig 12
An erratic analysis from LINGER.

The programs considered

The tentative conclusions that can be drawn from the evidence here are that there is a major division between the commercial programs and LINGER. This is not to say that LINGER is in any general or sustainable sense superior. As with all small programs it has the virtue of dealing with a small part of the data reasonably well but the vice of not being easily scaled up so as to continue to perform as well with larger data sets.

On the other hand, the commercial programs do not perform especially well. The principal reason for this is that they do not perform a full sentence analysis, rather they rely on various forms of pattren matching techniques which do not adequately capture the complexity of the data. There is probably nothing wrong with pattern matching *per se*. The patterns must, however, be sufficiently rich, consistent, comprehensive and above all share all the structurally similar featues with the 'object' they represent. This, none of the commercial programs do. Their 'objects' are simply words, with little that is especially linguistic between the word and the sentence. The linguistically crucial objects of the phrase and the clause are missing in most cases and it is their 'edges' that are vital in determining whether a subject and verb have in the first place been located and defined and in the second whether their relationship is grammatically correct - not the number of words in between.

Beyond a certain level of lexical pattern matching, it is not easy to see how these programs might be used to produce the kinds of rules and structures that LINGER has shown to be useful in identifying and diagnosing errors. At the same time, LINGER seems to have gone as far as it practically can.

Of all programs it has the virtue of dealing with a small part of the data reasonably well but the vice of not being easily scaled up so as to continue to perform as well with larger data sets.

Conclusion

As noted, all of these programs fail in terms of the criteria that have been used. They fail in different ways and offer different possibilities for extension or improvement. One question that has to be addressed is whether our evaluation of the programs is too exacting or critical. As with economics, language is by no means an exact science. 'Computing' language, however, suggests that in some respect our understanding of the structures and forms of language are, or can be, regarded as sufficiently exact as to be algorithmised, precisely defined and   processed in some comprehensive and consistent way. If this is so, such all-embracing analyses have not found their way into any of these   programs. It is unlikely that they will for some time.

More immediately, there are two major problems with the programs from the perspective of our target user. Firstly, despite claims of infallibility for the programs, the following shows that they are not:

> "*RightWriter* will not find all types of errors. In some cases, *RightWriter* will flag correct (or at least allowable) phrases as errors ... You must decide which of *RightWriter's* recommendations to follow."[26]

Far from identifying actual problems, conflicting claims of a high level of competence exist. Expectations are encouraged that cannot be fulfilled - *RightWriter* identified only about 10% of errors. An accountant would be highly critical of a spreadsheet that failed to perform a calculation with consistency, an information manager would be most impatient of a database program which failed to locate relevant records and an architect would not consider using a program that failed to translate drawings accurately into specifications of considerable detail. As users of language we should not accept programs that seem not to have a sufficiently high understanding of and respect for our profession.

Secondly, and more importantly, for the type of user identified here and even identified by one of the commercial programs, what is most worrying is the range of errors not identified, not only that the expectations are not always fulfilled but for the learner, there is no clear indication that the expectation of a complete check on a particular sentence has not been fulfilled. When one of these programs sails past an error without comment, then the user should be able to be confident in accepting that the sentence is grammatically unproblematical. However, if the user is a learner then there is a good chance that s/he will not know what has been missed. As self-help learning aids, therefore, all of the commercial programs have severe limitations, while LINGER is strictly limited.

[26] *Right Writer for Windows Guide* pp2-3, page 3.

References

Allen, J. 1987
Natural Language Understanding
Benjamin Cummings.

Barchan, J. 1987
Language Independent Grammatical Error Reporter (LINGER).
M.Phil. thesis. University of Exeter.

Boden, M. 1989
Computer Models of Mind.
Cambridge.

Bolt, P. 1988
An examination of the design principles underlying a computer
based English language learning environment.
M.Sc. Thesis

Bowyer, J. W. 1989
A Comparative Study of Three Writing Analysis Programs.
Literary and Linguistic Computing. Volume 4 Number 2. pp90-98.

Brown, K. and Miller, J. 1991
Syntax: A linguistic introduction to sentence structure. (2nd Edi-
tion).
Harper Collins.

Jackson, H. 1990
Grammar and Meaning: A semantic approach to English Grammar.
Longman.

Marr, D. 1982
*Vision: A Computational Investigation into the Human Representa-
tion and Processing of Visual Information.*
Freeman, San Francisco.

Obermier K.K. 1990
Natural Language Processing: an industry perspective.

O'Brien, P. 1989
eL: Using AI in CALL. in Yazdani, M. (1990) An English Tutor:
Project Report (1987-89).
University of Exeter.

Quirk, R and Greenbaum, S. 1990
A Student's Grammar of the English language.
Longman.

Reviewer's Notebook.
in *Byte*. September 1991 pp310-312.
Williams, N. 1991
The Computer, the Writer and the Learner.
Springer-Verlag.
Yazdani, M. 1990
An English Tutor: Project Report (1987-89).
University of Exeter.

CHAPTER 9

An artificial intelligence approach to second language teaching

Masoud Yazdani

Many Computer Assisted Language Learning (CALL) systems have been developed within the traditional Data Processing paradigm of computing. A recent display of such systems is presented by Cameron (1989). These systems are developed in order to help specific, and most focussed needs, of a particular teacher of a particular language. A system may help a learner with French noun gender (Farrington, 1986); another with English adverbs (Fox, 1986), and so on. Although the systems are motivated by practical pedagogical principles they do not have any "knowledge" of these principles.

As Cameron (1989) points out "the aim is not to show how ingenious we are in creating software but to use the computer to help us implement educational aims". This means that as with other forms of computer software the knowledge of the second language teaching is in the head of the programmer and not inside the system.

In contrast, a small group of researchers (Imlah and du Boulay, 1985; Barchan *et al.*, 1986; Pijls *et al.*, 1987; Yazdani and Uren, 1988; Cerri *et al.* 1989) have attempted to address the issue of language learning from the perspective of the Artificial Intelligence (AI) paradigm. As with most other AI workers, they feel that any effective form of human computer interaction requires both parties to have "knowledge".

The clearest exposition of the difference between the AI approach and the traditional approaches is presented by Brachman and Levasque (1985):

"What makes AI systems knowledge-based is not that it takes knowledge to write them, nor just that they behave as if they had knowledge, but rather that their architectures include explicit knowledge bases: more or less direct symbolic encoding of knowledge in the system."

Therefore, the starting points of someone addressing the CALL development, from an AI point of view is radically different from someone who is in need of a fix for a specific educational problem. While traditional CALL development is based on storage and application of massive patterns of potential interaction, the AI systems are based on simple but general principles.

Background

Frog (Imlah and du Boulay, 1985) aims to trap and comment on grammatical errors in an arbitrary French sentence. In its final form the program was only capable of parsing French declarative sentences. However, this system showed that a certain level of effectiveness can be offered without the use of anticipated situations.

FROG had a general knowledge of the vocabulary and grammar of French. However, when we tried to extend the system to broaden it beyond declarative sentences, we found it was not general enough in its architecture. The main problem we found with this system was that the knowledge and routines for processing the user's input were so intertwined that the change involved an architectural restructuring.

FGA (Barchan *et al.*, 1986) attempted to keep its "knowledge" of French grammar and dictionary separate from the processes needed to deal with the user's input. In addition, we added an explicit taxonomy of the common misconceptions of novice language learners (a "bug" catalogue).

While FGA had some success in achieving an architectural generality, it was still not effective enough in an everyday teaching situation. In the meantime, similar systems to FGA were developed for German and Italian which shared a good deal of similarity in their structure with FGA.

Barchan (1987) carried out a critical appraisal of FGA and attempted to build a language independent grammatical error reporter which could be used as the bases for systems in any European language. This system in fact has been extended to deal with Spanish (Yazdani and Uren, 1988) as well as English.

In 1988 we started the eL Project in order to develop an "intelligent" English grammar coach. The system was intended to be built in form of a general "shell" suitable for development of systems for those learners who are learning English as a second language.

From the outset, it was understood that the project would follow a "rapid prototyping" approach whereby the system would be tested by potential users whilst it was being developed. Due to the short time span of the project, only the first three months of each phase involved the preparation of a specification which was implemented, debugged and evaluated simultaneously during the remaining nine months.

In the first phase of the project we used the existing LINGER shell and developed the knowledge bases needed to make it work for English in addition to the original French knowledge bases.

As expected, the first phase of the project led to a second year, when the "first cut" prototype was re-implemented in PROLOG leading to an enhanced version of LINGER (O'Brien & Yazdani 1988; O'Brien 1992) . This new prototype (called eL) has been widely demonstrated and evaluated. An extra feature of the resulting system is greater concern with the user-interface design issues (Byron, 1990).

LINGER

LINGER, as well as enhanced LINGER (eL),was programmed in the logic programming language, PROLOG. The use of PROLOG as a tool shaped our view (Yazdani, 1989) of how one should build tutoring systems. Basically, we believe that the KNOWLEDGE and CONTROL should be separated. The "intelligence" in such systems comes from explicit representation of knowledge. These include knowledge of the subject matter (plus an inference engine); a bug catalogue (and user modeller); tutoring skills (and instructional planner) and explanation patterns (plus a student tutor interface).

LINGER is a "bug finder". The system would do its best to find any grammatical errors in a pupil's input. The diagram on the following page shows the way a sentence is processed by LINGER. LINGER, is general purpose, like an Expert System "shell". It can deal with any language for which it has appropriate databases. The user's input can be a free-form sentence in the language for which LINGER is configured. Originally, a French database was produced, with Italian, Spanish and German "toy" versions. Later, Bolt (1991) developed English databases for LINGER and Lawler (1990) extended the Spanish ones.

In order to extend LINGER's databases a user follows these 4 steps:

1. Define a minimal grammar of the language (say Spanish). This is the kind of thing a student might make as a crib sheet for a final exam. It might contain all the main rules of grammar (filtering out exceptions and minor cases) and provide a high level overview of the structure of the language.

2. Develop a project specific dictionary. This will have two separate bases ("a" and "b" below).

   a. most common and generally useful words in the language. This typically would be the collection of words that appear most often in speech and print (say the 500 most commonly appearing Spanish words).

   b. a more-or-less complete dictionary of words that pertain to some specific domain of activity (for instance, words that would be commonly used in a restaurant).

The way we expect that this portion of a dictionary will be generated is that someone will spend a lot of time in a restaurant listening to and recording what people say. We will build up an English vocabulary for the domain, then ask a native Spanish speaker to help develop the best Spanish equivalents.

3. Common mistakes: this is well covered in language teaching textbooks and taxonomy of the examples used by teachers of that language. One wants to look at the kinds of phrases that would be useful in the specific domain and then create perverted phrases based on commonplace phrases in restaurants.

4. Teacher explanations of errors: what is needed here is a fully articulate description of precisely what the error is from the point of view of a knowledgeable critic of usage.

Situated eL

In order to simplify the use and widen the audience of eL we plan to incorporate it into a more general learning environment. Pollard and Yazdani (1991) have begun this by building a multilingual (Spanish/French/English) restaurant scenario around eL using Hypermedia.

The package is intended for use in teaching language functions and vocabulary which are related to the interactions between restaurant staff and their customers.

From the outset we wanted the design to allow for the following:

- The same core program to be used for the learning of a number of target languages.
- Learner control of the interaction.
- The eventual provision of more than one mode of interaction, involving the learner to a greater or lesser extent in the active production of language. The most active learner role will involve the use of eL.
- More than one level of language difficulty.
- Linking the program to separate, independent dictionary lookup and/or grammar explanation facilities.
- Integration of graphics, sound and text.

An introductory screen presents the user with a number of choices as to how s/he wishes the session to continue. These choices include such things as the target language, the level of language difficulty, the level of passivity of the user and a number of settings which may influence the flow of the session.

The user then follows a route through a number of interactions between customer/s and restaurant staff in a typical restaurant scenario. There is more than one possible route, the particular route taken during a session depending on choices made by the user either on the introductory screen or during the restaurant dialogue. Thus, a user may use the program a number of times, achieving different results each time by making varied choices. A basic line of events, however, will have been followed, for there are a number of events which, are guaranteed to happen in a restaurant scenario (see Fig. 1).

Currently we have produced a program capable of working in 'demonstration' mode. In this mode, the user will be able to control the flow of the action through choice of icons at various stages. He will not, however, contribute actively to the actual language produced in the session.

The long term goal will be to allow the user to play the role of one of the characters in the scenario, entering this character's part of the dialogue in natural language. This input will be processed by a version of eL having some semantic content. It is planned to implementing more than the above two modes of interaction. These other modes

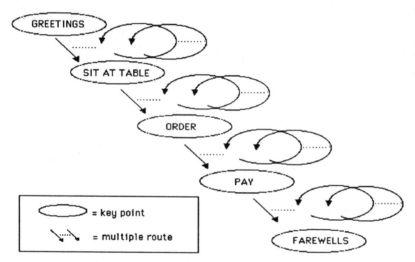

Fig 1
Flow of interaction
in a restaurant
scenario.

might allow the user to influence the language generated in a more general way than by entry of an entire dialogue item (making the water rude/polite etc.).

Related work

Pijls *et al.* (1987), have developed a prototype of an Author Environment which provides the user with automatic correction and explanation of grammatical errors and spelling errors of Dutch. Based on the same linguistic modules used in the Author Environment, the Nijmegen group has developed and tested knowledge based programs for grammar and spelling teaching, and an educational word processor with spelling and grammar checking for Dutch. These systems go beyond correction of the user's error by providing help and advice on grammar and spelling mistakes. The intention is to explain why a word is spelt the way it is instead of just correcting it.

Cerri (1989) has concentrated his efforts with ALICE on how the confusion of the meanings of words leads to mistakes made by novice language learners. Correspondence between words of two languages is seldom one-to-one and is highly context-dependent. ALICE currently deals with subordinate conjunctions in English, French and Italian. Schuster's (1986) system VP2 aims to explore how the "correspondence between the grammars of two languages provide an account of grammatical errors made by native speakers of one language attempting to learn a second language. "VP2 deals with two narrow bandwidths of the language; verb plus prepositional phrase such as "John ran into the street" and "Into the street ran John "in addition to verb plus particle such as"he filled up the bottle" and "he filled the bottle up".

While VP2 "maintains a model of the student's native language - Spanish. This grammar is assumed to be standard for all Spanish speakers." Also, the system would "fail if input contains unknown words or misspellings or unknown constructions outside the verb phrase".

Hypothesis

What happens when someone learns a new language? In order to build a computer system which attempts to be of some help we need to have an answer, however naive, to this question.

Our work has been based on a working hypothesis (Yazdani, 1990) which is clearly naive and possibly wrong. However, it would be useful to present this hypothesis as it may help the reader understand how our various projects fit together.

The hypothesis is as follows. When someone learns her mother tongue she does not build any explicit (high level) representation of that language, the learning of the language being linked to many other activities. Even if at any time some "introspective" knowledge of the language is developed, it is not needed in the everyday business of communicating with others.

When the person, normally at a later stage in her life, starts to learn a second language, she starts to become conscious of something called a"language". In learning a new language (L2) the learner attempts to make sense of this new language in the context of her mother tongue (L1). In so doing she builds a representation of what a language, any language (L*) is. She then tries to make sense of both her L1 and new L2 in the context of this general structure.

If we are to help a learner we need to consider that the learner is involved in a complex task. Not only is she learning specific facts about the structure of the target language (L2), but she is also rediscovering her mother tongue (L1). As she comes across more knowledge of the new languages he has to build a more complex L* structure. Accordingly, when she makes a mistake in the target language, she not only suffers from a confusion between her L1 and L2, but also she may suffer from restrictions or over generalisations of her L*.

This means that while Schuster (1986) is right in taking the mother tongue of the speakers in mind, she is wrong in assuming that all speakers of one language have the same understanding of their mother tongue, or suffer the same identical misconceptions in learning a new language. Whilst a population of learners with the same L1, learning the same L2, share some difficulties, they are also victims of their individual personal backgrounds. For example, someone who has learnt another language in the past would experience difficulties or advantages not shared by others.

Work in progress

An AI-based system which attempts to facilitate the learning of a second language within the above scenario needs to be both general and specific. General, insofar as it needs to deal with problems faced by learners, irrespective of of their tongue or the target language, and specific, in so far as it needs to have knowledge of the specific languages involved and the way in which interference may occur between them.

LINGER in the above sense is neither general nor specific. It does not have knowledge of any particular language nor does it have a general model of the user's learning process. We could spend many years writing a specification of a grand ideal design and then implement it from scratch. The one-shot approach (specification to implementation) of conventional software development is not an appropriate AI-oriented application. In our case the discovery of the specification is itself a major problem which can be solved by the use of a throw-away prototype such as LINGER.

The next stage of our work involves the design of a new system eL*
which uses eL-1 as a prototype. eL* attempts to use a model of the user's
mother tongue L1 in order to diagnose user's errors. In a collaborative
project with Pijlis *et al.* (1987) and Cerri's (1989) projects, we plan to
develop user models for Italian and Dutch people learning English.

The environment around eL* is to be provided by LE-MAIL, an
active form of email, involving some tutoring and monitoring. We
intend to encourage language learners to communicate with their pen
pals electronically. The user composes a message through a special
educational word processor, which incorporates a spelling checker
and which explains spelling rules, in addition to correcting mistakes.
The word processor is also used as the interface through which the
grammar analyser checks, offers advice and corrects users' grammatical errors.

LE-MAIL- a multilingual communication network

Another way of making eL situated would be to use it as a tool for users
of telecommunication networks to learn how to use a second language
without much effort (Yazdani, 1987). Such a system may be built as an
adjunct to the popular electronic mail services such as Eurocom, Inter-
net, Janet etc.

At the moment all email systems are passive carriers of information.
They expect the user to know what he/she wants to say. That means
they can be used between people who understand each other's
language. However, we propose LE-MAIL to be an active multilingual
carrier of information.

We propose a new alternative, LE-MAIL (Language Education -
Monitoring And Instructional Learning) as a multilingual communication facility which exploits powerful personal computers in order to
allow users to correspond with one another through existing telecommunication networks. LE-MAIL incorporates a suite of linguistic
systems which monitor the texts being created. It would detect spelling
and grammatical errors, offer specific remedial advice as well as
offering more general help with the grammar and spelling of the
language.

In this project we intend to encourage users to communicate with speakers of other languages electronically. The growth of the use of various Email systems indicates that our service would be attractive to many potential users. The difference here is that we shall support the user with various linguistic tools which allow the user to communicate with others on the network who may not know his/her language.

The user composes a message through a special educational wordprocessor. The processor incorporates a spelling checker which explains spelling rules in addition to correcting mistakes. The word processor is also used as the interface through which the grammar analyser checks, offers advice and corrects users' grammatical errors. The analyser would also incorporate help facilities. The advice will be provided using the knowledge of both the mother tongue and the target language. The system can be exploited both in school and commercial levels.

Reading tools (LE-MAIL-1)

As it is easier to write in one's mother tongue we propose that the user composes a message in his/her mother tongue which is transmitted to the friend. At this point the correspondent requires tools to understand the message.

Let us assume an English person sends the message and an Italian receives it. The Italian's personal computer can assist him/her to make sense of the message. Online dictionaries and grammar analysers which could be used as translation aids already exist and as the user will be in charge of the process the task would not be as complex as machine translation. Our system would not be capable of dealing automatically with the semantics of the messages and the receiver of the message would need to ask for clarification (from the pen pal) if the meaning of any message was not clear.

The Italian user of the system would reply to the message in Italian and in turn the English user has to make sense of the message using tools offered by his/her personal computer.

Writing tools (LE-MAIL-2)

The above scenario assumes that both parties to the correspondence take equal share of the work. However, it is possible that each user's computer may also incorporate (at a later stage of the project) tools which may assist him/her to compose messages in a foreign language. The writing tools would:

(a) detect spelling and grammatical errors,

(b) offer specific remedial advice

(c) offer more general help with the grammar and spelling of the language.

The intention is to explain why, instead of just correcting the mistake (as commercial wordprocessors do), we propose a development strategy which would allow a simple version of LE-MAIL with reading tools to be developed first, followed by its extension to include writing tools. Such a system could be used in a variety of application domains as is the case with the French Minitel system. As a starting point we propose that the system be aimed at the correct formulation of questions and answers about availability of actual information in another country.

For example a Fiat sales person in England may like to consult a Fiat manager in Italy about the availability of a certain model. The communication would naturally be within a restricted domain. In such areas the computers of both users can easily be provided with the complete dictionary of terms used as in the "restaurant" example above. Grammatical structures needed for such kinds of interaction also would be simpler allowing a grammar analyser to be constructed sooner than if one was to build an open ended system.

Acknowledgments

The work reported here is sponsored by the British Government through its training agency (formerly the Manpower Services Commission) and the Economic and Social Research Council (ESRC). I am grateful to my colleagues Keith Cameron, Paul O'Brien, Josephine Uren and Gordon Byron for their continuous support.

References

Barchan, J., Woodmansee, B.J. and Yazdani, M. 1986
A PROLOG-based Tool for French Grammar Analysis
Instructional Science vol. 14, pp. 21-48

Barchan, J. 1987
Language Independent Grammatical Error Reporter (LINGER)
M.Phil. thesis, University of Exeter

Bolt, B. 1991
eL: A Computer-based System for Parsing Corrected Written English
Computer Assisted Language Learning, vol. 4 pp. 173-182

Brachman, R.J. and Levesque, J. 1985
Readings in Knowledge representation
Morgan Kaufmann

Byron, G. 1990
eL: A Tool for Language Learning
Computer Assisted Language Learning, vol. 2 pp. 83-91

Cameron, K. (ed.) 1989
Computer Assisted Language Learning: Program Structure and Principles, Intellect/Ablex Publishing Corporation

Cerri, S. 1989
Acquisition of Linguistic items in the context of Examples
Instructional Science, vol. 18

Farrington, B. 1986
Computer Assisted Learning or Computer Inhibited Acquisition?
Cameron et al (eds.) *Computers and Modern Language Studies*
Ellis Horwood

Fox, J. 1986
Computer Assisted Reading - Work in progress at the University of East Anglia, Cameron *et al* (eds.) *Computers and Modern Language Studies*, Ellis Horwood

Imlah, W. and du Boulay, B. 1985
Robust Natural Language Parsing in Computer Assisted Language Instruction, *System*, vol. 13 pp. 136-147

Lawler, R. 1990
Dual Purpose Learning Environments
Computer Assisted Language Learning, vol. 4 pp.46-52

O'Brien, P. 1992
eL: Using AI in CALL
Intelligent Tutoring Media. vol. 3 pp. 3-21

O'Brien, P. and Yazdani, M. 1988
eL: A Prototype Tutoring System for English Grammar (Error Detector and Corrector), *Proceedings of the Third International Symposium on Computer and Information Sciences,*
NOVA Science Publishers, Inc.

Pollard D. and Yazdani M. 1991
A Multi-lingual Multimedia Restaurant Scenario
Interactive Multimedia, vol. 2 pp. 43-51

Pijls, F., Daelemans, W. and Kempen, G. 1987
Artificial Intelligence tools for grammar and spelling instruction
Instructional Science, vol. 16, pp. 319-336

Schuster, E. 1986
The role of native grammars in correcting errors in second language learning, *Computational Intelligence,* vol. 2.

Yazdani M. 1987
Artificial Intelligence for Tutoring
Tutoring and Monitoring Facilities for European Open Learning,
J. Whiting & D.A. Bell (eds.), Elsevier Science Publishers pp. 239-248

Yazdani, M. and Uren, J. 1988
Generalising language-tutoring systems: A French/Spanish case study, using LINGER
Instructional Science, vol. 17, pp. 179-188

Yazdani M. 1989
Language Tutoring with PROLOG
Computer Assisted Language Learning: Program Structure and Principles , K C Cameron (ed.), Intellect /Ablex pp. 101-111

Yazdani M. 1990
An Artificial Intelligence Approach to Second Language Learning
Journal of *Artificial Intelligence in Education* vol. 1 pp. 85-90

COMPUTERS, THE UNIVERSE AND EVERYTHING

An essential series of books looking at the way computers affect every aspect of modern life

Computers and Typography

Presented by Rosemary Sassoon

"No method ever affected our general education as deeply as the typographic method once did or as the computer is doing currently."
— Fernand Baudin

This book raises awareness of the most important issues in producing good typography and demonstrates why the lessons that have been learned in the past five centuries of traditional typography cannot be ignored. Indeed, the book itself stands as a model of good typographic practices. Contributions fall under five headings: Spacing and Layout; Typographic Choices – Latin and Other Alphabets; Technical Issues in Type Design; Lessons to be Learned from the History of Typography; and Research and the Perception of Type.

Price: £14.95, paperback   ISBN 1-871516-23-4

Computers and Language

Edited by Moira Monteith

The requirements of teaching in both English and Information Technology, together with the current and future demands of our society in which pupils will find work, mean inevitably that computers will be used more effectively and widely in schools. This book aims to encourage teachers and students to view the use of computers in language work in an enthusiastic way and brings together the wealth of knowledge and experience of leading international contributors.

Price: £14.95, paperback   ISBN 1-871516-27-7

Computers and Society

Edited by Colin Beardon and Diane Whitehouse

These papers represent an integration of research into technological
and social aspects of computer technology that is relevant and
accessible to the individual citizen. The book provides a discussion of
common societal objectives and provides a realistic assessment of the
role, and sometimes the limits, of computer technology in their
achievement. It will provide a reference point for future work on
computers and society in the 1990s.

Price: £19.95, paperback   ISBN 1-871516-41-2

Computers and Law

Edited by Indira Carr and Kate Williams

The use of computers in law has seen a rapid increase over the last
decade from simple access to legal information to the use of
computers for diagnostic and decision-making purposes. This has
resulted in an evaluation of the nature of law and legal reasoning.
This collection of articles covers the whole spectrum of computer developments
and aims to offer a synthesis of issues involved. Topics covered include Electronic
Data Interchange (EDI), criminal, intellectual property as well as evidential aspects.
It concludes with Artificial Intelligence applications to Law.

Price: £19.95, paperback   ISBN 1-871516-35-8

Computers and Creativity

Derek Partridge and Jon Rowe

This book is a study of human creative behaviour from a computational modelling
perspective. The authors examine theories and models of the creative process
in humans, both input creativity - the scientific, analytic side of devising
interpretations of input information - and output creativity - the artistic,
synthetic process of generating something novel and innovative. Topics
covered include: the nature and theories of creativity; computational
modelling; emergent-memory models; GENESIS; empirical
studies; and some conclusions.

Price: £14.95, paperback ISBN 1-871516-51-X

intellect European Studies

Humour and History, presented by Keith Cameron
A collection of papers which offer a fascinating insight into the role humour has played in various European cultures throughout their history.
£14.95 paperback   ISBN 1 871516 80 3   160pp

Children and Propaganda, Judith K Proud, Philip D Dine and Graeme Cook
Demonstrates how the literature of youth has been subverted to promote the ideologies of political regimes at key points in the 20th century.
£14.95 paperback   ISBN 1 871516 83 8 192pp Jan 1994

The European Community - Culture and Society, John Fletcher
£14.95 paperback   ISBN 1 871516 81 1 192pp Dec 1993

Regionalism in Europe: a study of regional aspirations in the Europe of the 90s
Peter Wagstaff
£14.95 paperback   ISBN 1 871516 84 6 192pp Dec 1993

Theatre and Europe (1957 to 1992)
Christopher J McCullough and Leslie du S Read
£14.95 paperback   ISBN 1 871516 82 X 192pp Dec 1993

intellect Computers, Language and Culture

The Art and Science of Handwriting,
Rosemary Sassoon
£19.95 hardback ISBN 1 871516 33 1 192pp

From Sumer to Jerusalem, John Sassoon
£14.95 paperback ISBN 1 871516 42 0 128pp

The Art and Science of Learning Languages,
Amorey Gethin and Eric Gunnemark
£19.95 hardback ISBN 1 871516 48 X 192pp

Antilinguistics, Amorey Gethin
£14.95 hardback ISBN 1 871516 00 5 288pp

Mac 3D, Stuart Mealing
£19.95 paperback ISBN 1 871516 46 3 240pp

Beyond Chaos, Robert Pepperell
£14.95 paperback ISBN 1 871516 45 5 192pp

The Linguistic Computation of Arabic,
Ajit Narayanan and Everhard Ditters
£19.95 paperback ISBN 1 871516 31 5 192pp

Mathematical Intelligent Learning Environments, Hyacinth Nwana
£19.95 paperback ISBN 1 871516 29 3 288pp

Representing Uncertain Knowledge,
Paul Krause and Dominic Clark
£14.95 paperback ISBN 1 871516 17 X 288pp

Inheritance Networks in Artificial Intelligence
Raad Al Asady and Ajit Narayanan
£19.95 paperback ISBN 1 871516 32 3 192pp

Reflections on Artificial Intelligence,
Blay Whitby
£14.95 paperback ISBN 1 871516 38 2 192pp